ICE MICE AND MEN

THE ISSUES FACING OUR FAR SOUTH

ANOTHER
OUR FAR
SOUTH
PUBLICATION

GEOFF SIMMONS & GARETH MORGAN
WITH JOHN McCRYSTAL

ICE MICE AND MEN

THE ISSUES FACING OUR FAR SOUTH

PiP
Public Interest Publishing

ACKNOWLEDGEMENTS

Visiting Our Far South is a once in a lifetime experience, but what made this trip extra special were the people aboard the Spirit of Enderby.

In particular we would like to thank the experts that accompanied us on the trip. They not only provided information and their expertise, but they also endured the painful process of thrashing out the key messages and transforming them into something that the public could understand. This process involved many late night conversations and hours spent reviewing the book, not to mention a few bottles of whisky. We also want to thank the organisations that allowed their staff to take part in this mission to raise awareness about the region. These and their organisations were: Rob Murdoch (NIWA – thanks also to all NIWA staff who contributed to this book), Lionel Carter & Dan Zwartz (Victoria University), Trevor Hughes (MFAT), Richard Levy (GNS), Andy Roberts (DOC), Gary Wilson (Otago University), Anton van Helden (Te Papa), Jack Fenaughty (Sanford) and Bob Zuur (WWF).

Thanks must go to all those who voyaged with us on the Spirit of Enderby, all of whom contributed to this book in their own way. The conversations and debates held on the boat helped us work out what issues were important, and how they were best communicated. So thanks for taking part, even when the Antarctic Circumpolar Current was playing havoc with your lunch. And for 'breaking the ice' and making the experience so enjoyable we would like to thank Nick Tansley, Rhian Salmon and Anton van Helden as well as Rodney Russ and the crew of the Spirit of Enderby.

Trying to present a holistic picture of the challenges facing a large chunk of the world is no simple task. Thanks have to go to John McCrystal for helping ensure that the issues were presented in a digestible fashion.

This project has been funded by the Morgan Foundation (www.morganfoundation. org.nz/) as part of its goal to stimulate debate on important public issues facing New Zealand. We don't ask that you agree with us on everything, but we do ask you to talk about it. Thanks to the Foundation for making the study possible. Geoff would also like to personally thank Gareth for providing the drive and vision to turn this dream into a reality.

Finally all errors and omissions are of course our own and none of those mentioned above can be held accountable for the views expressed. If everyone is equally upset with our views, then we have probably got it about right.

Cover Image: David White, Mark Fraser, Rob Murdoch, Ray Hoare.

Designed by typeface ltd.
Printed in Wellington, New Zealand by Printlink.
First published in 2012 by The Public Interest Publishing Company Ltd (PiP).

Enquiries to Phantom House Publishing:
Fax: +64 4 384 5451
Email: info@phantomhouse.com
Web: www.phantomhouse.com

ISBN 978 0 9876666 2 8

CONTENTS

INTRODUCTION

I (Gareth) have been privileged enough to make a once in a lifetime trip to Antarctica, twice. In 2008 I went to Antarctica with a group of mates on a voyage that opened my eyes to the wonder of the area south of New Zealand. Joanne and I spent the whole trip marvelling at the teeming life of the subantarctic islands, the fury of the Southern Ocean and the twisted, chiselled white shapes of Antarctica.

When we got back it felt weird – we had been to places that were part of New Zealand, yet no one back home knew about them. I wanted to talk about what I had seen, but most people couldn't understand. Young Kiwis now regularly go to Africa to see the big game as part of their OE, yet they don't realise the incredible wildlife on our own back doorstep. It just seemed wrong.

Over the next few years I started to notice all the issues that were facing this part of the world. It's like when you buy a new car and you start noticing other people driving the same car. Every time Antarctica or the subantarctics came up, my ears would prick up. I'd had an introduction to the impacts of climate change on melting ice in Antarctica and changing currents in the Southern Ocean as we researched *Poles Apart*. For *Hook, Line and Blinkers* we learned about the threats posed by fishing, pollution and plastic to the Southern Ocean. Then Malcolm Templeton's book *A Wise Adventure* opened my eyes to the unique way Antarctica has been managed by nations working together. This part of the world is not only stunning and important, it is facing huge challenges from a changing world.

It all seemed to come together into one hair-brained idea. Why not go back? And this time, why not take some experts with us so that the Kiwi passengers can learn about the issues facing the region. In fact, what the hell, why don't we try to take *all* Kiwis along with us for the ride? We could use the voyage to raise awareness about the region and the issues facing it. I wanted people to know about New Zealand's far south, appreciate it and the special role it plays in the planet's politics, climate and wildlife.

I figured if the edu-vacation idea was going to have any legs then other organisations with an interest in *Our Far South* would have to buy into my vision and come on board; literally. They did, and before I knew it the trip was a goer. Antarctica New Zealand, the organisation that coordinates our presence on the White Continent, was first to sign up. The National Institute of Water and Atmospheric Research (NIWA), Ministry of Foreign Affairs and Trade (MFAT), Department of Conservation (DOC), Victoria University and the Institute of Geological and Nuclear Sciences (GNS) soon joined them. Initially I thought they might send the tea lady as their 'expert', but it slowly dawned on me that we had collected New Zealand's top experts on one boat! Over the following months Te Papa, Sanford and World Wildlife Fund (WWF) also put reps on board. I didn't want zealots – these people had to spend a month on a boat together without killing each other – but nor did I want moribund pointy heads. We had all the right ingredients for sparks to fly, but hopefully without setting fire to anything.

In February 2012 ten experts, forty ordinary Kiwis and one bear mascot (named after the explorer Shackleton) set off on an adventure we called *Our Far South*. On that voyage we got to rub shoulders with New Zealand's top experts on everything from the Antarctic Treaty to marine mammals to the impacts of melting ice sheets on sea level. This book is the result of that intense month of presentations, experiments and late night whisky-fuelled debates on the boat. It begins with a look at why *Our Far South* is so important to New Zealand and the planet. We then look at the issues facing the region; including whether there is a race for resources, the impacts of climate change, and the challenges facing the region's precious wildlife. And to help you on your way we have included a colour section for images in the centre of the book – these will be referred to in the text.

WHY IS OUR FAR SOUTH IMPORTANT?

Welcome to *Our Far South*. Imagine you are standing at Bluff Point. You are about to board a boat heading south. What will you see? What natural forces will determine the experience you will have – winds, seas, temperatures, precipitation? Why would you go south — there's nobody there, is there? Should we really care about this part of the world?

We all know about Stewart Island, but how many of us know about New Zealand's territory beyond that? *Our Far South* is unique, and despite the virtual absence of human inhabitants, it is one of the most important areas in the world, not just because of its unique wildlife, but because of its role driving the world's climate. Many of us wouldn't have a clue that our own backyard is so critical globally. As we will show, the natural environment of *Our Far South* (that's what we'll call it) is intricately interconnected to other environments thousands of miles north, where human occupation does dominate the landscape, although it's connected in ways we are only beginning to understand. Consequently we don't fully know what the impacts of humankind's actions will be in the region, either from our race to exploit its natural resources or its reaction to the unfolding drama as we fail to arrest climate change. But it's not all bad news. *Our Far South* is also a rare example of successful governance by the international community — a unique geopolitical achievement that has bestowed any number of benefits, but which faces its own challenges.

This book will look at a number of the issues that face the area south of Stewart Island, an area where New Zealand's interest stretches all the way to the South Pole, where the apex of our triangular Ross Dependency meets the end of the earth (see Image #1). Is an uncontrolled race for resources already underway, or can we be confident that we have created the institutions needed to protect the region? What impact will climate change have and, in turn, how will that affect us in New Zealand? Is the wildlife of the subantarctic islands, the Southern Ocean and Antarctica under threat, and if so, should we care?

Before we look at these issues we really need to understand how the oceans, the climate and the ecosystems of *Our Far South* function and why the overall system is just so important from a global perspective. Like most environments, its various elements are interconnected, and it connects in various, intricate ways with the world at large. Understanding these connections is crucial to understanding how a simple change or shock can echo across the entire area and even reverberate globally. In this book, we'll take you on a metaphorical journey south, voyaging through three different zones, each defined by unique combinations of geology, oceanography, climate and biology. The difference is that our lens will take a closer look at the things that can't easily be seen, showing how everything works together — and why we should care about the region. The other advantage of a metaphorical journey is that you can't get metaphorically seasick.

First, consider the subantarctic islands. The waters around these islands are generally warm enough that they never freeze. This helps keep the islands free of ice, ensuring they are a valuable refuge for wildlife. Hence they are often known as 'liferafts' in the Southern Ocean.

The second zone is the Southern Ocean itself. This zone is delineated by the Subtropical Front (STF) to the north and extends south to the ice shelves and glaciers of Antarctica. The

STF is where the warm, subtropical waters mix with the cooler waters from Antarctica, as can be seen in Image #2. The flow of this STF is interrupted by the South Island; it curls around the bottom and spins up the coast as far as Christchurch before veering all the way out along the Chatham Rise and then on to South America. Because nutrient-rich cold waters from the south mix with the iron-rich warmer waters of the north, it is around the STF that a lot of plankton blooms happen. Plankton is the base of the marine food chain, so more plankton means more life in the ocean, and more fish and chips for us (okay, just fish).

The Southern Ocean really gets going where the continental shelf around New Zealand's subantarctic islands drops away to the deep abyss below. Here, the westerly winds and currents are uninterrupted by land so they can flow freely around Antarctica, building up to tremendous speeds — the only place in the world where this happens. In fact, this is the world's largest and longest current system, known as the Antarctic Circumpolar Current (ACC — any resemblance to our compulsory accident insurance scheme is purely coincidental, although you may end up on ACC after sailing through the ACC because these are the roughest seas on Earth). This region is also the engine room of the global ocean, and of the world's climate. That is what many of us don't realise and in our ignorance we're complacent about the changes it is undergoing.

The third and final zone as we venture south is Antarctica, including the sea ice that surrounds it. It is too cold in the centre of Antarctica for life to flourish; most wildlife is in the water or on the coast surrounding the continent. This zone is dominated by ice – the slow flow of glaciers to the ocean, and the annual pulse of sea ice formation and melt. Every year Antarctica doubles in size thanks to the efflorescence of sea ice (that means to bloom like a flower, not that it glows in the dark), and this process helps drive the marine food chain for Antarctica's wildlife. The annual freeze and thaw of ice, along with wind, drives the world's global ocean circulation which transports heat, gases and the nutrients essential for marine life right around the world.

Zone One: Subantarctic Islands

Our subantarctic islands appear to be isolated, but as can be seen on Image #3 they are all actually part of a single huge, submerged continent which oceanographers refer to as Zealandia. This stunning topography is a result of some serious plate tectonics. Plates are the rigid slabs of crust on which sit continents and islands and that move about slowly on a fluid substrate like bits of bread dropped onto cheese fondue. For millions of years, Antarctica was part of a mega-continent called Gondwana, along with New Zealand, Australia, India, Africa and South America. Up to 170 million years ago, this mega-continent shared the same plants and animals.

Zealandia is a fragment of ancient Gondwana that eventually sheared away and was submerged. The collision of the vast Pacific and Australian tectonic plates eventually forced up a bunch of islands along the boundary: these are now collectively known as New Zealand. The effect of this fall and rise of the land was a bit like sheep-dipping New Zealand's islands (though not necessarily all at the same time). This swift (in geological terms, anyway) dunking could explain why we lost many of the land-based animals and plants that we shared with the other fragments of Gondwana (except for a few, such as the humble tuatara) and the land was re-colonised by birds.[1]

The United Nations Environment Program dubbed our subantarctic islands 'the most diverse and extensive of all subantarctic archipelagos'. All five island groups were honoured with UNESCO World Heritage status in 1998. Nearly as prestigiously, they are also National Nature Reserves under New Zealand's Reserves Act 1977. The Department of

Conservation is charged with protecting and preserving these islands in perpetuity.

We Kiwis mostly don't appreciate what a set of jewels we have lying in the ocean to *Our Far South*. They may not be a tropical paradise, but they are full of so much life and are such a hotspot of endemism (that is, so many of the creatures that live here are unique to the area) that they have been dubbed the 'Southern Serengeti' or the 'Galapagos of the South'. The five island groups are:

- Bounty Islands
- Antipodes Islands
- Snares Islands
- Auckland Islands, and
- Campbell Island.

When we took the physical journey south from which this book drew its inspiration, we also took in Macquarie Island which, while politically part of Australia, is geologically and ecologically part of our own subantarctics.

The five island groups (six if you include the Australian-owned Macquarie) that make up the subantarctic islands are located in the Southern Ocean south to south-east of New Zealand (for a map of the region, see Image #1). The groups that lie in our territory span six degrees of latitude, from 47 to 52 degrees south, in other words occupying the stormy latitudes of the Roaring Forties and Furious Fifties, known also as the Albatross Latitudes. Macquarie Island lies a bit to the south-west, and (at 54 degrees) further south.

There are other subantarctic islands in the Atlantic and Indian Oceans which you may have heard of, such as South Georgia and Kerguelen Islands. However, our subantarctics are the only islands in this part of the world. They are home to plants and animals found nowhere else on the planet, and act like vital 'liferafts' for many of the creatures that live in the south. Among the species that live there are albatrosses, penguins, spectacular flowering 'megaherbs', parakeets, invertebrates (stuff without spines like insects and politicians), snipe, sea lions, and elephant seals. They also represent the southern limit of the stunning southern rata.

How can these islands support so much life? They are pretty tiny, and don't exactly offer much in the way of things to eat. Most of the animals in *Our Far South* live off the bounty of the ocean. The subantarctic islands are not important so much as a food source as they are critical as a refuge for those animals that need to come ashore to rest, breed and raise their young. Once their breeding is over, they travel huge distances to forage for food. It is usual for the seabirds that shelter here to traverse vast tracts of the globe on the back of the southern westerly winds: for example, shearwaters even go all the way to Alaska and back. On average, our New Zealand sea lions swim over 200 km on each foraging trip, and are the deepest diving 'eared' seal in the world.[2] More on these critters later.

Before humans came along, these islands were packed with seals and seabirds. But we decimated the indigenous populations of many of them and replaced them with our usual entourage of pigs, cattle, rabbits, cats and sheep, as well unintended hangers-on such as rats, mice and weeds. All of this laid waste to the delicate ecosystem of the islands. The 19th and early 20th centuries saw a massive slaughter of seals, whales and seabirds, from which many populations have never fully recovered.

The subantarctic islands seemed to lose their value once they were stripped of seals, and only served as a menace to shipping. But the slaughter of the animals that lived there and the aborted attempts at farming and settling the islands was part of the process that allowed them to be claimed as New Zealand territory. And ironically, they and their surrounding waters now combine to give New Zealand an Exclusive Economic Zone

(EEZ) of over 4 million km², which is in the top five largest in the world. International law gives New Zealand the sole prerogative to decide how to use the ocean resources within this area of ocean. On this basis, we can argue that New Zealand is an oceanic superpower, laying claim to a submerged territory that is larger than India. And because we own the massive submerged continent of Zealandia, we also have a claim over the seabed right to the edge of the continental shelf. This takes us to within 280km of the 60th parallel – the start of Antarctic waters and the Ross Dependency, which is yet another huge slice of land and ocean under New Zealand's stewardship. The fascinating question is how we wish to use that power. Do we want to use it to protect the environment, or to exploit the area's bounty, or both? New Zealanders have some choices to make.

Zone Two: Southern Ocean

The feature that really makes *Our Far South* unique is not Antarctica lurking about at the bottom of it; it's the ocean that surrounds it. As we've noted, this is where practically all of the wildlife of the subantarctics have their livelihoods. And as we've hinted, the Southern Ocean is the engine room of the global climate.

Some 34 million years ago, Australia and Zealandia separated from Antarctica, and along with a mobile South America created a passage of deep water all the way around the Southern Hemisphere. The opening of this last gap, between the tip of South America and the Antarctic Peninsula (known as the Drake Passage) allowed the westerly winds and currents an unimpeded romp around the globe. This accident of geography created the world's greatest current system – the ACC. And it was the inauguration of the ACC that directly contributed to a massive shift in the Earth's climate from hot to cold, which we will detail below.

The ACC truly is a flow of superlatives. At 24,000km long, it is the world's longest current, and the only major flow to connect all the major oceans. It is also the largest current in the world by volume, pumping almost 150 million cubic metres of water every second – that is the equivalent of almost 300 Amazon rivers! As we will see, this current is driven by all the wind in the Southern Ocean; in fact, the ACC soaks up about 70% of the wind energy falling upon the total global ocean surface.[3] That is a ridiculously high percentage if you think about it for a moment and look at a map of the world – this current only operates in a small fraction of the world's oceans, and yet it absorbs the bulk of the wind energy that those oceans are exposed to. The power of this wind energy is plain for all to see: the Southern Ocean is renowned for having the biggest waves in the world. In the Pacific, waves from storms in the Southern Ocean have turned up a few days later as good surf in California.[4] Little wonder this current is so influential – and it is in our back yard.

In fact, the ACC is a major determinant of the entire world climate. When it formed 34 million years ago, global climate cooled dramatically and the first ice sheets appeared on Antarctica as the ACC started to draw down carbon dioxide (CO_2) from the atmosphere (we'll look at how this happens shortly). Even today, the world's oceans soak up 30-40% of the carbon emissions from human activity, and they have also absorbed 90% of the extra heat generated by human-induced warming, slowing the pace of climate change (for now). The Southern Ocean is the biggest player in this absorption of heat and carbon dioxide, and it is because of the ACC.[5] If this current hadn't formed, things would be an awful lot hotter right now.

More detail on what creates the ACC is available in the box below. Then we'll move on to focus on what a difference the ACC system makes to the planet.

A Bad Case of Wind

Being the one place on Earth without much land, the Southern Ocean is where winds and currents can have their wicked way. It all starts with wind – the scourge of our beloved capital Wellington (although it does keep the smog at bay). These winds are created by two forces – the heating and cooling of air (known as *convection*) and the spin of the planet (known as *the Coriolis Force*).

Convection is the result of differences in temperature. The sun beats down on the Equator, and not so much on the Poles. This heats up the air (and water), causing it to expand, get lighter, and rise. Once it rises, the air starts to cool and any water it contains condenses to form clouds; as it rises, it leaves behind (near the surface) less air – known as a patch of 'low pressure'. You might know this from the big 'L's you see on your TV weather maps. These warm air masses eventually cool and descend, forming areas of high pressure (again, our friend the 'H' from telly). Meanwhile, as nature abhors a vacuum, air generally rushes from areas of high pressure to areas of low pressure, creating wind. Generally speaking, this means the tropics have low air pressure and the poles have high air pressure. So the cool air from the poles rushes towards the high pressure at the equator, replacing the warm air from the equator with cool air from the poles. This helps smooth out our planet's temperatures. The larger the temperature difference between the equator (hot) and Pole (cold), the stronger the winds are as the air tries to reduce the difference. Pretty cool, huh? (excuse the pun).

In the temperate regions, where New Zealand and the Southern Ocean sit (between 30-60° south), at surface level warm air is moving south from the equator to the pole. This gets turned into a westerly wind by the *Coriolis Force* (created by the Earth's spin). When air is at the equator, it picks up some extra momentum from the spin of the Earth. When it moves south, it still carries that momentum, but now the Earth isn't spinning that fast. It is a bit like jumping from a moving train. If you jump right at the point where you want to get off, your body will retain momentum from the moving train and you will hit the ground beyond the point you were aiming for. So it is with air: the momentum gathered at the equator turns a south-bound wind into a westerly wind.

It's because of the lack of land in the Southern Ocean that the westerly wind has free rein, and that's why we have the 'Roaring Forties', and their even stronger southern neighbours, the 'Furious Fifties' and 'Screaming Sixties'. This is why trees on the West Coast grow with a permanent lean to the east. The winds make our weather so dynamic and difficult to predict that people come from all over the world to study it. For meteorologists, predicting the weather in New Zealand is the equivalent of working on the New York Stock Exchange. Yet farmers still grumble when the experts get it wrong.

This westerly wind whips the ocean up into the very powerful Antarctic Circumpolar Current (ACC – see Image #4 for a map of the current). You can think of the ACC as a bit like a bunch of kids playing in an old-style, circular swimming pool. Normally they splash and play and there is no real current. But if they all move in the same direction for long enough, they can work the water up into a strong current. Even a bunch of spindly 8-year-olds can do it, and then spend the next half-hour being dragged around the pool by the force of the current they've created.

The ACC is extremely strong. There is evidence that in the Southern Ocean, the wind creates currents that are felt all the way to the sea floor, four kilometres below! That interaction with the seabed is critical. If the ACC did not 'feel' the seabed, it would spin out of control. The ACC needs the seabed to produce friction that balances the effects of the winds at the surface. That friction takes the form of sea floor erosion and movement of sediment.

Antarctic Circumpolar Current (ACC)

The Earth's climate has always gone through natural fluctuations. The way in which the Earth orbits around the Sun and the way in which it spins and wobbles on its own axis are a major factor in this: you can see the impact of these subtle variations in the records of ice sheets; over natural cycles lasting up to 100,000 years, they grow and shrink as regularly as if it were the pulse of the planet. These variations are known as Milankovic cycles, after the mathematician who first described them. The planet's temperature can also fluctuate with periods of more intense solar activity (which are heralded by sunspots) and decrease with the large-scale emission of particles such as ash and sulphur dioxide (from volcanic eruptions) into the atmosphere. But these changes typically last a few years compared to the big Milankovic swings and roundabouts.

When the ACC came along, however, there was a massive shift. In the distant past, back when dinosaurs roamed the Earth, the planet was still going through these business-as-usual cycles of cooling and warming, but on average it was 6-7°C warmer than it is now. In this period, known as the 'Greenhouse Earth' phase of our planet, there were no permanent ice sheets on land, and atmospheric concentrations of CO_2 were much higher than they are now, at around 1000 parts per million. Around 34 million years ago, permanent ice sheets started appearing on Antarctica, which signalled the start of our current climate phase, known as 'Icehouse Earth'. Part of the reason for this change was the creation of the ACC, when South America and Australia separated from Antarctica. So while you might curse the ACC when you are being buffeted by a nasty southerly gale surging straight from Antarctica itself, it is quite important to us. Without it, we would have a much warmer planet, which means higher sea levels, and less land – and frankly, it's quite likely we wouldn't exist.

The whirling ACC also exports water throughout the world's oceans, providing 75% of the water for nutrient-rich upwellings that occur in other parts of the world, helping to feed the world's fish.[6]

All of the benefits of the Southern Ocean come from the mixing or churning (or to use the scientific term, 'overturning') of water that happens thanks to the ACC. It isn't crucial to understand *how* the current mixes water, but if you are interested, it is set out in the box below. Then we will look at why the mixing role performed by the ACC is so important for the planet.

Stirring the Pot

Without wind and currents, water will sit in layers with warm water on the surface. The Southern Ocean is different, thanks to the westerly wind created by convection and Coriolis Force as we saw above.

When the wind hits the water, the Coriolis force comes into play again, and this proves to be crucial in influencing the ACC system. When the wind hits the water it is westerly: in other words, it comes from the west. This pushes the water towards the east, but under the influence of the Coriolis effect, the water is directed to the north. This shifts surface water from the edge of Antarctica and moves it north into the main oceans. This water then sinks to the middle depths of the ocean (around 1400 m), carrying heat and carbon dioxide with it. This is a big deal, as it helps to reduce the impacts of human influenced warming.

Deep water from the northern Atlantic, Pacific and Indian Oceans returns to Antarctica (see Image #5). Some of it rises to the surface to replace the Coriolis-driven water heading north, and the remainder moves under the Antarctic ice shelves and begins to melt them from beneath. In doing so, it becomes a super-cold – 2°C, but doesn't freeze because it contains the extra salt that is left behind by the water that forms ice in winter. This super-cold, salty cocktail is very heavy and sinks to the very bottom of the ocean before heading to the north.

Thus the Southern Ocean has two ways of mixing the ocean (wind and ice), and between them they move water north and south in each of the major oceans. The combined effect of this is that the Southern Ocean mixes the ocean more effectively than any other area on Earth.

The other effect of this upwelling is that most surface water swirling around in the ACC never quite makes it to Antarctica. This helps buffer Antarctica from temperature changes in the Southern Ocean. The ACC acts like the walls of a chillybin, helping keep the contents of Antarctica cool. This isn't impermeable: some water always sneaks through in swirling eddies or gyres (the main gyres are in the Ross and Weddell Seas).

ACC Storing Heat

Ocean currents actually do the lion's share of the work redistributing heat around the globe. There are a couple reasons why this is the case. First, most land can't help much with rapid convection – that movement in the saucepan when you boil the porridge – because the molecules in it are fixed in one place. By contrast, water is fluid, so the heat can move around. Water is also transparent, so radiation can penetrate more deeply than it can the solid ground. That is why the surface of land masses heats up and cools down relatively quickly and doesn't store heat that well. You see this with the huge temperature swings you get inland compared to the coast. In some inland deserts, the temperature can vary by 50°C in a day, reaching the maximum temperature in the midday sun before dropping to zero during the night. On the coast, the adjacent water absorbs and releases heat more slowly, keeping the temperature more constant.

So that leaves air. While there is a lot of air around and it is transparent to incoming solar radiation, it's quite insubstantial stuff and cannot hold much energy; all up, water is about 1000 times denser than air. In fact the top 3.2 metres of the ocean holds as much heat as the entire atmosphere![7]

This ability of water to absorb and redistribute heat allows the ocean to even out the extreme highs and lows provided by the seasons and helps shift heat around the planet. The ACC is so important to the world's climate because of its mixing of waters (due to high winds), which carries warmer surface waters into the depths and presents colder water to the warming effect of the sun. This means more heat can be absorbed by the ocean, and it can be absorbed more quickly, especially with the other currents that spin off the ACC into all the world's oceans. Hence, the Southern Ocean truly is the engine room of the world's climate.

ACC Storing Carbon

This mixing also conveniently stores away carbon dioxide in the deep ocean. Carbon dioxide has been building in the atmosphere since humans started burning fossil fuels. Scientists have noticed that there is less carbon dioxide in the atmosphere than there should be, given the amount of fossil fuels we have burned – indeed, about half of the CO_2 produced from emissions has 'disappeared'. There are only two other places the carbon dioxide can go – terrestrial ecosystems or the ocean. Terrestrial ecosystems haven't had the chance to absorb much; how could they? We've been busy cutting trees down.

It turns out that the ocean is the biggest store of CO_2. It actually holds over 50 times more carbon than the atmosphere[8] and appears to have absorbed about 40% of the CO_2 that humans have already emitted. Like heat, the lion's share of this has been stored in the Southern Ocean. Because the deep ocean is so damn big (the Pacific Ocean alone can accommodate the entire world's landmass), it is far and away the largest store of carbon dioxide on our planet.[9]

The ocean recycles and stores carbon dioxide in two ways – dissolving it in seawater (known as the *physical pump*) and collecting corpses of carbon-based marine critters on the seafloor (known as the *biological pump*). We'll take a quick look at the biological pump before we focus on the physical pump.

The process of the biological pump is similar to that by which carbon gets locked up in soil on land. Living things containing carbon die, and rot. Much of their carbon component is released back into the atmosphere in the process of decomposition, but depending on the conditions, some of those carbon compounds get turned into soil. Similarly, in the ocean, some creatures die and either rot en route or reach and form a deposit on the seafloor. Their corpses contain carbon, either in their bodies or in their shells made of calcium carbonate. Some of this carbon can become trapped at the bottom of the sea in sediment or, in the case of calcium carbonate, as a limestone deposit (which over time can get shunted up by earthquakes to make things like the Waitomo Caves or the White Cliffs of Dover).

Unfortunately most of it doesn't stay locked away in Davy Jones' Locker. Thanks to the nutrient cycling we will see later, most dead stuff gets broken down so that it can be reused. So while the biological pump does result in some large deposits of carbon being stored away on the seafloor, it is not the major driver of the amount of carbon stored in the ocean, which brings us to the so-called 'physical pump'.

Most of the missing CO_2 has simply been absorbed by the ocean. It turns out that carbon dioxide mixes with water pretty well. As a result, the world's oceans play a huge role in absorbing carbon dioxide from the atmosphere (and therefore controlling the climate), especially as human emissions of the greenhouse gas continue to rise.

The ocean is huge and could potentially absorb a lot of carbon dioxide, but two factors control the pace at which this happens: ocean mixing and temperature. Most oceans of the world don't mix much, so only the surface layer of the ocean is in contact with the atmosphere. As we humans have increased our carbon emissions, the surface layer of the ocean has become saturated with carbon dioxide – it simply can't absorb any more. Second, warm water can absorb less gas. You can test this yourself by filling two glasses with a fizzy drink, leaving one in the fridge and keeping the other out on a warm day. The warmer one will go flat more quickly because more carbon dioxide will bubble out of the drink than the cold one, where a proportion of the gas remains dissolved.

In the Southern Ocean, these limits don't apply. Thanks to the ACC, it's really windy and rough, and all the mixing that happens as a result increases the amount of water that has contact with the air. The water is also cold, so it can absorb more carbon dioxide. That is why 40% of the carbon stored in the ocean is taken in between 30 degrees south and 50 degrees south. A key component is what happens in the northern boundary of the ACC, where gas-rich surface waters sink (or *subduct*), carrying heat and gases to the ocean's interior. At the other end of the world, the ocean around Greenland is also sinking and carrying carbon dioxide-bearing water to the deep ocean.

The Southern Ocean provides an incredible service (as we economists put it) and without it, we would have been feeling the impacts of climate change much sooner than we have. But that service comes at a price: the oceans are becoming more acidic, and this is predicted to continue over the next century. Seawater is naturally basic (the opposite of acidic), but as CO_2 dissolves in water, a complex series of chemical reactions ensues that makes the solution slightly more acidic. We all know acid eats away at things: in the ocean, it reduces the amount of carbonate in the ocean available for marine organisms such as plankton and corals to build their exoskeletons and shells. There is growing evidence that the oceans are acidifying more rapidly than they have for millions of years, particularly the Southern Ocean.

ACC Nutrient Cycling

The ACC's mixing also plays a major role in the biology of the ocean, helping to get nutrients back up to the surface of the ocean where the sun is. Nutrients, sun, air, chlorophyll and something that knows how to use it – these are the ingredients for photosynthesis.

Like life on land, life in the ocean (particularly plant plankton, known as *phyto*plankton or algae, which are the base of the food chain) needs light and certain basic nutrients to grow. The main nutrients needed are carbon, nitrate, phosphate, silicate, oxygen, hydrogen and sulphur. A few other trace elements are also critical in very small quantities, such as iron, but more on that later. There is only a limited quantity of these nutrients around, and so to keep the circle of life turning, we rely on bacteria to take waste materials or dead things and turn them back into nutrients for algae to use. This is called nutrient cycling. As we saw in our book on fishing *Hook, Line and Blinkers,* nutrient cycling is one of the most important roles of the ocean and the seafloor.

The trouble is that when things die in the ocean, they tend to sink. Bacteria can break down the nutrients for algae to use again, but the hard part is then getting those nutrients back to the surface of the ocean where the sunlight is so that plankton can grow. This is where the ACC comes in. The water in the lower parts of the ocean is enriched with nutrients such as nitrogen, phosphorous and silica. The mixing caused by the ACC helps stir these nutrients up and bring them back to the surface. Then it redistributes those nutrient-rich waters throughout the Southern Ocean and beyond.

For this reason, the waters of the Southern Ocean lapping at New Zealand's southern shores are the most nutrient rich on the planet, as can be seen in Image #6.

As we have seen, this current system whizzes around Antarctica, connecting all the world's oceans, and the strength of this current is such that it radiates sub-currents all over the globe; like a poorly maintained ride at a fair, this spinning occasionally sends bits and pieces flying off in all directions. In so doing, the ACC supplies nutrients to most of the rest of the world's ocean, too. Here and there, the nutrient rich water eventually rises to the surface and allows plankton to bloom. All in all, an incredible 75% of the world's fisheries are fed on nutrients that originate in the vortex of the ACC. The box below shows that New Zealand is one of the beneficiaries of these side currents, and how our proximity to that side current could have contributed to our name *Aotearoa* – the land of the long white cloud.

Land of the Long Plankton Bloom?

It may not just be our mountain ranges that help make New Zealand such a cloudy place. Incredibly, when algae grow, they emit a chemical that some scientists think could have a massive impact on our planet's climate.

Water mixes up with nutrients in the swirling ACC around Antarctica. The ACC is so strong it generates large eddies (swirls) when it collides with the southernmost part of Zealandia – the great Campbell Plateau. Six to eight times a year, these giant vortices stir the ocean from top to bottom even where the water is deeper than 4000 m, almost like a deep-ocean storm. At the ocean surface, this water moves north to the Chatham Rise, where it meets warmer subtropical waters from the north and forms the Subtropical Front. The southern water is rich with nutrients, which mix with the essential minerals like iron and zinc that are introduced by rivers or that can be blown all the way from Australia in a dust storm. Along with sunlight, this heady cocktail provides all the vital ingredients for algae (technically known as phytoplankton, the tiny plants that most ocean life relies on for food) to bloom. This, in turn, gets the whole oceanic food chain going, feeding fish like our massive hoki stocks.

As a result of photosynthesis, algae emit a small number of chemicals. One of these chemicals is Dimethyl Sulphide (DMS), the same chemical that creates the boiled cabbage smell when you, well, boil cabbage. Some researchers think that seabirds might even be able to pick up that boiled cabbage smell from high up in the air and use it to zero in on the potential food source growing in the ocean far below. Scientists have so far failed to explain how they avoid crashing into the houses of Russian immigrants preparing their traditional *shchi* – cabbage soup.

Each plant only emits a tiny amount of this chemical, but when you get a massive plankton bloom, the gas can end up being created in reasonably large quantities. Once in the atmosphere, DMS breaks apart and combines with oxygen to form new chemicals such as sulphuric acid. These chemicals can act as 'seeds' for water vapour to gather around, forming clouds. So it may well be that productive oceans contribute to our reputation as Aotearoa, the Land of the Long White Cloud.

Given the amount of nitrate in the waters of the Southern Ocean, you'd expect it to have the most life of all oceans on the planet, but this isn't the case. This is thought to be due to the comparative lack of light, the lack of iron (which is due to the lack of land, from which most iron originates as dust), and the excessive churn of the ocean, which quickly sweeps the algae to depths where light is so low that it retards photosynthesis. But never fear: despite this lack of life, the nutrient rich waters are not wasted! They rise to the surface in other places all over the globe, helping to grow plankton and feed the fish there.

So to the casual outside observer, the Southern Ocean might appear to be just a churning, sloshing mass of waves; the kind of place you would curse if you were to sail through it (!). But look a little deeper, and you'll see that the unique, colossal ACC régime of shallow and deep currents impacts on the world's climate, the available nutrients in all the major oceans and the amount of carbon dioxide in the atmosphere. Have a good think about the cooling benefits from that sucking up of heat and carbon from the atmosphere while you're throwing up over the side.

Zone Three: Antarctica

The ancient Greek 'scientist' (there were no PhDs back then) Ptolemy (who lived in the 1st century AD), believed that symmetry required there to be a landmass at the bottom of the Earth in order to 'balance out' the rest of the world's landmasses. This idea permeated much of the ancient world, but it took nearly another 1700 years until any actual attempts were made to prove the theory true.

Ancient cartographers dubbed this mythical continent *Terra Australis Incognita* (it simply means 'Unknown Southern Land', but it sounds cooler in Latin). The fallacy that a Great Southern Continent were needed to balance things out has long ago been exposed: in the big picture of Earth floating in space, the highest mountains and deepest ocean trenches on the earth's surface are mere pimples and dimples. However, as we've described, it turns out that having a continent on the Pole surrounded by ocean is still pretty important.

The English explorer, Captain James Cook, is credited with the first close encounter with this southernmost continent. Historians agree that Captain Cook was the first to cross the Antarctic Circle and came within 75 nautical miles of the continent on his voyages in 1773 and 1774, but while he suspected his proximity, he never knew how close he really came.

People finally sighted the Great White Continent in 1820, although historians dispute who got there first. What we do know is that in 1841 the British naval officer James Clark Ross, who had already discovered the North Magnetic Pole, explored the Ross Sea (see Image #1) in search of the South Magnetic Pole. This area is now the centre of New Zealand's claim over Antarctica, known in his honour as the Ross Dependency. (Note: the Ross Sea proper is not to be confused with the 'Ross Sea region' which includes the Ross Sea but extends north to 60 degrees South). The Ross Sea was the starting point for the most well-known Antarctic event: Amundsen and Scott's race to the geographical South Pole in 1911-12.

Yet for all the intrigue and mystery that surrounded its existence, and then its nature, the Great Southern Continent ended up being lumbered with the rather hum-drum name 'Antarctica', which simply means 'Opposite the Arctic.' Factually speaking, Antarctica does indeed lie at the bottom of the world, opposite the Arctic. But it is a little like naming your child 'junior'. The truth is that Antarctica has some significant differences with the Arctic. These differences have big implications for climate, biology and geopolitics, as can be seen from the box below.

Arctic vs Antarctic

The Arctic and Antarctic are often mentioned together, but they couldn't be more different. The Arctic is mostly ocean surrounded by continents, whereas Antarctica is a large continent covering 10% of the world's land surface, surrounded by the swirling Southern Ocean. The Arctic has historically been covered in sea ice, which tended to stick around year after year because it is hemmed in by the continents around it. By contrast, the current around Antarctica causes most of the sea ice to disappear each year in the summer.

These differences have a huge impact on the climates of the northern and southern hemispheres. In winter, both places are cold, but in summer the continents around the Arctic warm up quickly, raising temperatures. This reduces the temperature difference between the equator and the North Pole, calming the wind and currents. This generally gives the Northern Hemisphere strong, definite seasons – a freezing, stormy winter followed by a balmy, calm summer.

By contrast, Antarctica is covered in two massive ice sheets, both kilometres thick. This ice makes Antarctica the coldest place on the planet, with winter temperatures of −80 to −90°C (your average freezer is around −18°C, so fill your mouth with ice cream straight out of the freezer then multiply the ice cream headache by 4-5 times). Thanks to this huge block of ice and the insulating effect of the ACC, temperatures remain fairly cold even under the midnight sun of summer. So even in summer, the Southern Hemisphere has a big temperature difference between the equator and pole. This means we have less definite seasons, and get storms and wind year round. It also means that there's always the chance of a polar blast making its way all the way to Northland. Oh, goody.

And of course, the Arctic has polar bears and the Antarctic has penguins, who never meet each other in the wild (luckily for the penguins).

Antarctica is the coldest, windiest, driest, and (on average) the highest continent in the world. Antarctica holds about 90% of the entire world's ice, which in turn comprises about 70% of the world's entire fresh water supply. That means that the ice on Antarctica holds more than twice the amount of water that you can find on land, in the air and in all the de-humidifiers in the entire world. There is enough ice there to cover the full extent of Australia to a depth of 3.5km, which might sound like a great idea come Bledisloe Cup time.

Antarctica is divided by a huge range of mountains known as the Trans-Antarctic mountains (see Image #1). The tallest peak in this range is Vinson Massif, which at almost 4,900 metres would tower over our own biggest peak, Aoraki/ Mount Cook, at 3,800 m high. At one end this mountain range juts out into the ocean, forming the Antarctic Peninsula. The area west of the mountain range (closest to New Zealand) is smaller, and known as West Antarctica. At the centre of West Antarctica is the Ross Ice Shelf, which is a huge mass of floating ice as large as France at the southern end of the Ross Sea. The other side of the mountain range is known as East Antarctica, which makes up the bulk of the White Continent.

Having such a high, thick ice sheet over the South Pole creates many unique climatic features. For example, Antarctica has winds that have been recorded at up to 327 km/h; much faster even than the winds in a Category 5 hurricane (like the winds in Hurricane Katrina or Rita). This is caused by air that is cooled by the elevated regions of the cold continent and then according to the forces of gravity rushes down the mountainsides of Antarctica towards the coast, creating 'katabatic' winds. As we will see, these winds have a big impact on sea ice formation.

Ice

In the same way as the Southern Ocean is dominated by the ACC, Antarctica is dominated by ice.

Antarctica is technically a desert because of the very low level of rain (or snow) that falls there. However, because Antarctica stays so cold even in summer, the snow doesn't melt much and has built up over time and compacted to form a vast sheet of ice. Gravity makes it flow slowly downhill to the sea, at just a few metres per year. Near the edge, the flow comes together in much faster glaciers and ice streams, which can move several hundred metres per year. All in all, snow falling in the deepest interior parts of the Antarctic ice sheet can take over 100,000 years to reach the ocean. Still, given their size, that amount of movement is actually quite fast, and the result is that beneath the glacier or ice sheet, the lower surface of the ice carves up the landscape like an almighty grinder. U-shaped valleys such as those you'll find in Fiordland were once carved by such glaciers: look at the rock walls and the gouging action of the ice is written there for all to see.

Despite the low snowfall, the continent still manages to accumulate around 2,000 billion tonnes of ice per year. As a result, Antarctica's ice sheet is, on average, around 1.8 km deep. That is more than five Auckland Sky Towers piled one on top of the other. The sheet over East Antarctica is larger and deeper, at 2.3km average depth compared to West Antarctica at 1.3km. At its deepest, in East Antarctica, the ice sheet is over 4 kilometres deep, as can be seen in the cross section of the Antarctic Ice Sheet below.[10] Incredibly, the sheer weight of this ice is enough to depress the crust beneath Antarctica by between 0.5-1km.[11]

Cross Section of Antarctica

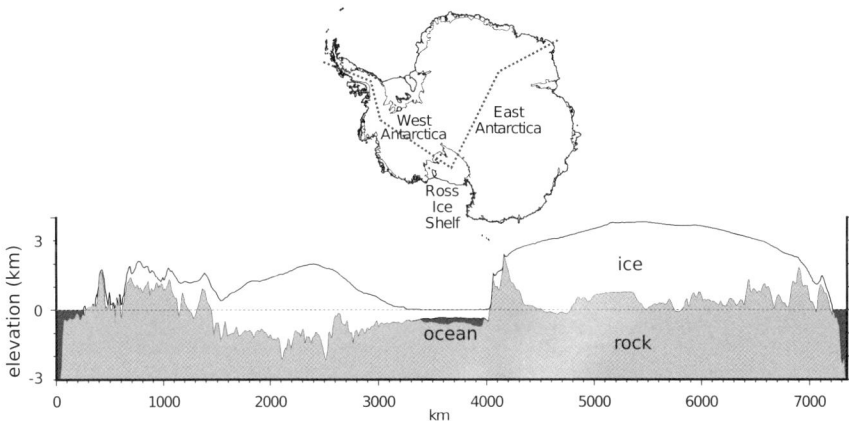

Image courtesy of Dan Zwartz. The cross section follows the line on the map above.

The massive, permanent ice sheet covering Antarctica gives it the appearance of being a solid continental land mass when it isn't. The majority of Western Antarctica is not actually as solid as it looks; big chunks of it are below sea level, as you can see in the cross section above. The reality is that Antarctica, particularly in the west, is not so much a Continent as an archipelago of islands connected below sea level, not unlike New Zealand! Some of this land would rise if there weren't so much ice on top of it, but West Antarctica would still mostly be below sea level.

Antarctica is so cold that ice almost never melts, and glaciers flow all the way to the ocean. Once the ice reaches the ocean, the glaciers can form an ice tongue that floats on top of the ocean, and this eventually breaks off to form icebergs. If there are a bunch of glaciers or an ice sheet entering the same bay, they can spectacularly join up to form great ice shelves. The two largest ice shelves are the Ronne-Filchner Ice Shelf in the Weddell Sea and our very own Ross Ice Shelf in the Ross Sea. Ice shelves make up around 11% of the total area of Antarctica. The Ross Ice Shelf is almost twice the area of New Zealand (NZ is 268,000 km^2 and the Ross Ice Shelf is 483,000 km^2) and is several hundred metres thick. That is one big popsicle!

Ice also forms on the surface of the ocean each winter. This is how it happens. During the winter the pole is tilted away from the sun, and so stays in complete darkness for 24 hours a day. The air temperature drops well below zero as a result, while the ocean tends to stay a frosty – 1.8°C year round. So when the surface of the sea starts freezing, it sometimes seems to smoke, as the water evaporating at the surface immediately turns to ice crystals in the air. Then the surface of the water becomes very heavy-looking as a thin skin of ice forms (known as 'grease ice'). Eventually the ocean develops a thin layer of fresh ice known as *nilas*. When wind comes, the *nilas* breaks up and the little bits bump into each other, creating the very photogenic pancake ice. Eventually these pancakes pile on top of each other (known as rafting) and over winter a thicker layer of sea ice forms, usually around 1.5 m thick. Voilà, we have sea ice! If that ice is this season's ice, it is known as one-year ice. But if it is left over from last season or even older seasons, and built upon by subsequent freezes, it is known as multi-year ice, which is very thick and hard and hazardous to the health of shipping.

The sea ice rises and falls with the tides, so where the sea ice meets the edge of the land there is often a break or two which allows it to move up and down with the tide. In summer the pole tilts towards the sun, causing the sun to shine for 24 hours a day. This raises the temperature, and works with the current and winds around Antarctica to break out and melt most of the sea ice.

The annual sea ice expansion is one of the world's great natural events. The area of sea that freezes is the same as the Antarctic continent itself, so it appears to double in size. That is a whole extra continent just made of ice, albeit pretty thin. At 14 million km^2 Antarctica is already twice the size of Australia, and with all that ice it becomes almost as large as Africa. In summer, the sea ice around Antarctica almost disappears, shrinking to about 20% of its winter extent.

Like any good infomercial, Antarctica even throws in extra ice for free! Antarctica is so efficient at creating ice, it even exports it. We'll explain. Remember those katabatic winds that scream down the mountainsides in Antarctica? Well, when they hit the coast, they push any newly formed sea ice or icebergs out to sea. Where the ice pack is forced away from the land in this way, it forms a polynya – a Russian word for 'window' used to refer to a semi-permanent gap in the sea ice where new ice keeps forming but gets pushed away from the coast by the wind. This makes these polynya areas hugely productive creators of sea ice.

This whole chilly business is incredibly important to the world's climate, for several reasons. Having all that ice sitting on top of the Antarctic continent means that an equivalent volume of water is not in the ocean – which means that sea levels are some 65 metres lower than they could be if the world's ice all melted. The sheer expanse of all that white snow and ice also plays a significant role in moderating our climate through what is known the *albedo* effect (actually nothing to do with your sex drive). Albedo is the amount of light (and therefore incoming solar radiation) that is reflected by an object. During the six months it receives sunlight, Antarctica reflects a considerable amount of radiation and back into space which would otherwise warm the surface. The albedo effect therefore maintains global temperatures lower than they would otherwise be.

However, perhaps the most important role of the ice in Antarctica is the contribution it makes to deep currents, and to the Antarctic ecosystem as we will see later.

Ice and Deep Currents

We have already seen why ocean mixing is very important. You'll recall that to the north of Antarctica the ACC and the Coriolis force causes water to upwell. Some of that water heads north and is mixed by the ACC to the middle depths of the ocean. The rest of it heads south to Antarctica, where it ends up sinking to the very bottom of the ocean, as we shall now see.

Ocean water near Antarctica is both very cold and salty, both factors which make the water very dense, causing a massive *downwelling* – sending tonnes of water to the bottom of the ocean. This sets off the world's entire system of deep currents. How does this happen? Water circulating beneath the ice shelves becomes super-cold – down to − 2°C. It can actually become colder than the freezing point of water because of the presence of salt in the water. Water with salt (or any dissolved solid) in it is harder to freeze, (or boil) because it has more mass to cool down (or heat up). The cold makes the water dense, and therefore heavier. As water freezes to form sea ice on the surface it leaves behind its salt, which makes the remaining water even heavier.

So the water circulating beneath the great floating ice shelves becomes both saltier and colder than normal water. Both these factors mean that large quantities of water get denser and heavier than normal seawater, and sink to the bottom to depths exceeding 3.5 km. These downwelling water masses spread out along the bottom, spilling through the oceanic trenches and troughs and displacing other water masses, and driving the world's deep currents.

So the ACC and sea ice come together in an almighty natural symphony, as shown by Image #5. The icy bit on the picture represents the Antarctic coastline: in other words, the right of the picture is actually south. The Antarctic Circumpolar Current (depicted by the yellow lines) whips from west to east as well as pushing water to the north under the influence of the Coriolis force. This leaves space for deep, nutrient-rich water from the North Atlantic to surface, recycling nutrients that would otherwise be stuck in the deep ocean. Some of this water heads north and sinks to middle depths. The rest heads to Antarctica and is made colder and saltier as it circulates below ice shelves and sea ice to be converted to Antarctic Bottom Water. The way salt and water temperature interact to affect the density or buoyancy of ocean waters is known as the *thermohaline* effect.

The deep, cold salty water then sets off on a merry Kontiki tour through all the world's oceans, as we saw in our book on fishing *Hook, Line and Blinkers*. This is not to give the impression that density is the only driver of this circulation. As we have seen with the ACC, wind is a big driver, especially for surface currents. Furthermore, more attention is turning to the role of turbulence in keeping this global circulation going. Like everything we face in nature, it becomes more complicated the more research is undertaken. There is some truth to that cliché that 'we know more about the surface of Mars than we do about Earth's oceans'. Which is no surprise, really, as Mars has no ocean.[12]

We need to be thankful for these currents. They slowly draw a huge amount of heat and carbon dioxide into the deep oceans, regulating our climate. They are also responsible for the nutrients in the ocean, which ultimately affects whether you can get any fish to go with your chips.

Ice and the Antarctic Ecosystem

The other crucial role of ice shelves and sea ice is that which it plays within the Antarctic ecosystem. Of course, the ice provides a significant habitat for creatures living both on and under it. But by far the most dramatic impact of sea ice comes when it melts.

As we have seen, the waters of the Southern Ocean are nutrient-rich. All that is missing for plankton to grow is sunlight, reduced churn of the water and some essential elements such as iron. When spring and then summer roll around, all these conditions are met and there is an explosion of life. The 24-hour sunlight melts the sea ice, which allows light to penetrate the water. The fresh layer of melted water temporarily cuts off the deep water circulation, which reduces churn and provides more settled conditions for plankton to grow. And the melting sea ice even manages to provide iron. We don't entirely know why, but one theory is that wind-blown dust containing iron and other essential elements (perhaps from places like the Dry Valleys in the Ross Dependency) settles on the ice and then is made available to the plankton when the ice melts. Another theory is that as the glaciers move down to the ocean, they take up the sediment underneath, which contains traces of iron that they then dump in the sea.

So for a short period of time when the sun shines and the sea ice melts, the chilly Antarctic waters become the most productive ocean on the planet. Then, for the rest of the year, they return to their usual fallow state. This boom and bust of plankton supports a unique ecosystem.

This pattern began with the isolation of Antarctica some 35 million years ago, with the opening up of the ACC. It's no coincidence that at the same time we saw the evolution of whales split into two modern groups: baleen whales and echo-locating toothed whales. Baleen whales stopped bothering to grow teeth, which had allowed them to catch prey one by one, and started growing baleen (the bristly curtains in their mouths), which allowed them to filter small creatures from the ocean in huge numbers. This was to prove a vital adaptation to the Antarctic environment.

The boom and bust cycle of life in Antarctica favoured predators that could handle the cold, which means they have to be warm-blooded like whales, penguins and seals, or have antifreeze in their blood like toothfish. They also need to be able to pig out for a short space of time then live off their fat reserves for the rest of the year. Through the process of evolution, this is a niche that whales, particularly large ones like the blue whale, became uniquely positioned to fill. With their massive mouth size, they can scoop up enormous quantities of krill (a small, shrimp-like creature) in a single gulp. During these few months of feeding frenzy, they can build up the fat that allows them to migrate to the tropics for breeding, and back for next year's banquet. During this migration and the breeding season, they eat virtually nothing. Must have other things on their mind.

Antarctica was thought to be unique in that it had one of the shortest food webs in the world. All food webs start with plants, but with most of the ocean's top predators you usually have to go through four to five other stages before you get to a plant. For example a killer whale might eat a big fish, that ate a little fish, that ate a prawn, that ate a tiny zooplankton, and that ate some algae (a plant). This is important, because it turns out that eating is a very wasteful process — not just because of all the bones and icky bits that predators leave behind, but because each animal only gains about 10% of the energy mass of the critter it devours. But in Antarctica, algae is eaten by krill, which in turn is directly on the menu for whales, fish, seals and birds. This is a very short food chain, and a very efficient way to feed very big creatures like whales and seals. As with everything there are exceptions, but this works as a general rule.

Krill are often fingered as the critical link in the Antarctic food chain. They occur in incredible quantities, which can only be measured in millions of tonnes. Krill live in tightly-packed swarms (upwards of 30,000 individuals per square metre) that rise and fall in the ocean's water columns. A single swarm can contain up to 2 million tons of krill and can span 450 square kilometres. The total amount of krill can vary hugely, but some guesstimates put the historical amount of Antarctic krill at some 500 million tonnes. That is five times the total amount of fish caught in the whole world every year, making krill one of the largest protein sources on the planet.

Krill are certainly dominant around most of Antarctica, except in the Ross Sea where the silverfish is dominant. This seems to be because the main krill species, *euphasia superba,* lives in the open ocean, whereas the silverfish are better adapted to the shallower and more southern waters. Sea ice is essential in the life cycles of both krill and silverfish. Silverfish can live under sea ice all year around, while krill need to hatch and survive under sea ice during winter, while they wait for the spring and summer bonanza of algae. As you can imagine, food is pretty hard to come by during these dark winter months.

Incredibly, while the winter storms rage above the sea ice, the temperature in the water underneath remains relatively stable, at just below zero. As mentioned, salt allows the water temperature to drop slightly below zero, but it can go no further or it will turn to ice. This means that underneath the sea ice is a very stable environment for the life living there, at least compared to the environment above the ice. In places where the ice doesn't melt, this means it has the same temperature and calm conditions all year around. So even in the meagre hours of sunlight that autumn and early spring provide in Antarctica, algae are able to grow under the sea ice. It is a bit like a lawn, but upside down. This algae is what nourishes krill and silverfish during the tough winter months and, if necessary, silverfish can live off them all year round.

Incidentally, when the sea ice melts this same algae is released, and contributes to the bloom that kickstarts the whole Antarctic food chain. All in all, sea ice is just as critical a habitat for the critters living beneath it as it is for the cute penguins and seals we see diving off or huddled on top of it.

Summary

This is just a taste of the wonders of *Our Far South*. It is a place of incalculable importance to New Zealand and to the entire world. The ecosystem, climate and the actions of humankind are irrevocably intertwined – maybe here more than anywhere else on the planet.

Our journey south started in the subantarctics, where the cool, nutrient rich waters of the south combine with the mineral-rich warmer waters from the north, creating plankton blooms. In this area, the ocean's bounty provides for an incredible quantity and diversity of wildlife. Much of this wildlife relies on the subantarctic islands for refuge and breeding, like liferafts in the vast, empty ocean.

Next we found ourselves in the incredible, churning Southern Ocean. While it does not carry much life in itself, the unimpeded romp of currents and winds around the globe allows the ocean to mix in a way that isn't possible anywhere else. The churn caused by the Antarctic Circumpolar Current (ACC) means the Southern Ocean stores more heat and carbon dioxide than any other region and exports nutrients all around the world. The current may not be much fun to sail through (depending on your tastes), but it has profound impacts on the world's climate and the life in the world's oceans.

Finally we reached the ice of Antarctica. The ice of Antarctica works in tandem with the ACC to drive the world's ocean mixing. So having a continent on our southern pole, surrounded by ocean and carrying an immense quantity of ice is part of what makes our planet's current climate so hospitable. The centre of the continent is too inhospitable for life, but at the edges, particularly in the zone where the incredible pulse of the annual freeze and thaw cycle of sea ice happens, there are short bursts of life in amazing quantities. This boom and bust ecosystem gives rise to a completely unique arrangement of life.

This overview opens a window to the majesty of *Our Far South*. Throughout the rest of this book, we will look at the issues and challenges confronting this remarkable region. We'll begin by looking at whether there is pressure from a race for resources in *Our Far South*. Then we will move on to the effects of climate change and how those effects might echo throughout the world. Finally we will see how all these pressures are impacting on the wildlife of *Our Far South*, and what this could mean for their future.

SECTION TWO:
A RACE FOR RESOURCES?

*[Antarctica] is the last beautiful, vast, virgin land in the world…
to go it alone is pure madness." – Roald Amundsen*

Who controls a piece of land? Or for that matter the ocean? Throughout history, the short answer has been that might is right. Whoever has the biggest gun gets their way.

To the *conquistadors*, seeking riches seemed like a welcome relief from their miserable existence farming the thankless *meseta* (arid, high plain) of Western Spain. They felt they had God's personal authority to conquer the New World, so they did. To ease their consciences, they read out a remarkable 'legal' document – known as the *Requerimiento*. This ordered the native people to submit to Spanish rule and allow themselves to be preached to by Catholic missionaries. If they resisted, they were held fully responsible for any of the ensuing consequences, such as war, slavery or death. Deal or no deal? Most bizarre of all, the *Requerimiento* was often read out in Spanish to an uncomprehending local people, or sometimes even from the deck of a ship before an invading party even landed.

This Spanish hubris proved to be short-lived; one hundred years later, Protestantism was rife, the Pope's word was worthless and bits of vessels from Spain and Portugal's joint Armada were washing up on British beaches. Without the steel to back up their claims, the Portuguese and Spanish territories withered just like all Empires before them.

It took until the twentieth century before international law got much more sophisticated than that. By the nineteenth century, European countries were really hitting their strides, claiming most of Africa and a good portion of Asia to boot – just because might was right.

What did it take to claim sovereignty, aside from the biggest gun? Over time, international law developed some agreed principles to tone down the somewhat bestial edict of might being right. Sovereignty usually required occupation – which is made up of *animus* (intent to possess the territory) and *factum* (actual occupation, or at least use of the resources in the territory). Contrary to popular belief, the discovery of a territory alone was considered to be a good start, but usually not enough. You couldn't just plant a flag and say job done. In a nutshell, sovereignty was first-come, first-served, then use it or lose it.

The unfortunate side effect of this was that it encouraged a 'race for resources'. Whoever could find new land or resources and exploit them the fastest got to keep them. Of course, all too often by the time this 'race for resources' ended, there were no resources left – they'd been pillaged, leaving little more than a wasteland. Strip mining of phosphate begun by the British in Nauru is a great example of rendering a territory barren. Has this happened in *Our Far South*, and more importantly, is it still happening?

The Race for the Subantarctics

The early history of the subantarctic islands is very much a race for resources. There's a silver lining — it left New Zealand in charge — but the wildlife of the islands paid a high price for our adventurism.

There is archaeological evidence that some of the subantarctic islands were discovered by some adventurous Polynesian sailors: middens and charcoal found on Enderby Island in the Aucklands group indicates that the island was periodically visited many hundreds of years before Europeans passed that way. There doesn't seem to have been any attempt to settle there, however: after that, there was no human presence until boats started sailing past on their way from Europe to the colonies in the Pacific (there was no Panama Canal until World War One). The first of New Zealand's subantarctic island groups was discovered in 1788, when the *Bounty* (of the mutiny fame) happened across the eponymous bunch of rocks in the Southern Ocean while on its fateful mission to pick up breadfruit in the tropics. Captain William Bligh commented that he saw patches of snow on the barren rocky outcrops, but while it gets easily cold enough for snow, this was more likely guano – bird poo.

The remainder of our subantarctic islands were discovered in the space of the next 22 years, and the race for resources started forthwith. The sealers were first in, eager to cash in on the bonanza of seal skins and oil and leave nothing for their competitors. Government attempts to manage the plunder failed, as the islands were simply too remote to prevent poachers. As a result, seal populations were decimated. Between 1780-1833, seven million fur seals were killed in the Southern Hemisphere, leaving so few that the industry became unprofitable.[13] The population has since rebounded somewhat, but thankfully sealing has never re-commenced in the Southern Ocean.

Macquarie Island even had a plant for rendering oil, which ironically was run by erstwhile mayor of Invercargill, Joseph Hatch. This plant originally processed elephant seals, but Hatch turned his attention to the superabundance of penguins after they'd all but exterminated the elephant seal population. Each penguin — Royals were favoured, but as they began to run out, Kings were used instead — yielded about a pint (600 ml) of fine, light oil, and there were millions upon millions of penguins around (at first). The operation was always marginal, not least because Hatch could never quite seem to find his way clear to properly provisioning, paying or even sending a ship to pick up his men as promised. Hatch's oiling licences were eventually revoked by the Tasmanian government, largely at the petition of a public movement championed by the Australian polar explorer, Douglas Mawson: this is widely considered to have been the world's first wildlife conservation campaign.

Meanwhile thanks to the Antarctic Circumpolar Current, the strong winds and currents in the Southern Ocean had made the area a virtual motorway for sailing ships trying to get back and forth from the colonies. The importance of the 'Great Circle Route' turned our subantarctic islands into the equivalent of a broken bottle right in the middle of the road; the rocky outcrops made it the most dangerous shipping route in the world. Many shipwrecks ensued — nine in the Auckland Islands alone. The New Zealand Government put depots on the islands and liberated livestock so that any starving shipwrecked sailors didn't have to live off sea lions. Meanwhile, of course, the livestock liberated — pigs, goats, rabbits — lived off the indigenous flora of the islands, while their inadvertently introduced counterparts — rats, cats, mice and even dogs — preyed on the fauna. They were still there when the Age of Sail ended and the Panama Canal opened in 1914, rendering the Great Circle Route obsolete.

The rather bleak Auckland Islands and Campbell Island were even eyed up as sites for farming and as whaling depots. In 1849, a British whaling firm was awarded a royal charter and sent 150 people to establish a settlement in Erebus Cove on Auckland Island. When they arrived at Port Ross they found 40 Māori and 30 of their Moriori slaves were already there, having fled from the Chatham Islands after they had killed some French whalers. Other Māori settlements had also sprung up on other parts of the islands. The British settlers founded the town of Hardwicke but it proved to be short lived. They had come prepared to grow the usual range of English crops, but none (like wheat) could grow in the harsh conditions. The fleet of whaling vessels they expected to send business their way had all but disappeared along with the whales, and a little over two years after it was founded, the settlement was abandoned. The Māori settlements lasted a little longer, subsisting on seals and flax, before they too chucked in the towel.

The New Zealand Government offered a pastoral lease over Enderby Island from 1896, and there were a couple of attempts to make it a goer. All failed, but the sheep, cattle and cats remained. A lease was also offered for Campbell Island, and it was cleared and farmed (unprofitably) until the 1930s. Bureaucrats in Wellington even tried to get someone to run a farm on the Bounty Islands. Noticing they had an unused patch of land on their map, the New Zealand Government ran an advert for someone to take up a 335-acre pastoral leasehold on the rocky islands. Someone clearly overlooked the fact that there was no vegetation on the islands — well, a single clump of Cook's scurvy-grass was spotted there a few years ago — as they frequently get swamped by the massive Southern Ocean and are consequently little more than a set of guano-encrusted rocks.

In the end, in every case, the brutal wind and rain of the Southern Ocean won the day; even rugged Kiwi farmers couldn't turn a living from these islands. And no wonder: Campbell Island gets winds of around 100 km/hr one day in three, and it rains (or snows) on average six days a week. Average temperatures range from 5-11°C through the different islands, making Southland look positively tropical.

This race for resources damaged the local wildlife to such a degree that some of the islands and their plants and animals have yet to fully recover. But some good did come of it. Under international law, Britain (on behalf of its colonies New Zealand and Tasmania) was able to claim sovereignty over the subantarctic islands. At the time, no one in the international community was interested enough to object – after all, they were merely tiny rocky outcrops in a vast ocean, with little prospect of useful settlement. In fact, for a while they seemed to cause more trouble than good: New Zealand spent a lot of time and resources rescuing stranded sailors from the islands. New Zealand activity on Macquarie Island led to a dispute with Tasmania over who really had sovereignty over the islands, which was eventually ruled in the Aussies favour. As the Phoenix Yellow Fever like to chant: "same old Aussies, always cheating."

And so the situation remained until 1982, when the United Nations Convention on the Law of the Sea (UNCLOS) came along, and the subantarctic islands delivered a massive windfall to New Zealand, as explained in the box below.

UNCLOS

It sounds more like a 1960s secret agency, but UNCLOS is actually a really important agreement over how nations manage the sea and its resources. Before UNCLOS came along, management of the ocean was still a reflection of the old might is right days. A nation's control over the ocean only extended three miles – the distance that you could fire a cannon from shore. But beyond this, the oceans were viewed by all as *mare liberum* – literally, the free seas. This principle was based on the notion that no power could claim sovereignty over the ocean.

At the start of the twentieth century, nations started realising that there was gold in them thar waves. Some countries started making noises about rights over the ocean in order to control pollution, safeguard fish stocks and invest in mineral or oil extraction. Technology meant that the oceans' resources were being accessed in new ways, and at greater depths: in 1947, the first offshore oil platform was operating. It became clear that without some rules, it would be a free-for-all and conflict would surely result. In 1945, the United States seized upon their new-found superpower status to claim their continental shelf, securing the rights to oil that existed there. This prompted a goldrush – many countries followed suit, claiming continental shelf extending a variety of distances depending on the resources they wanted to secure – fishing, mining or oil. As is the way with international negotiations which generally require consensus, it took until 1982 to agree a solution.

UNCLOS created the 200-mile 'Exclusive Economic Zone' (EEZ), which gives countries rights to exploit the natural resources within 200 miles of their coast. In exchange, other nations have full rights of thoroughfare by ships and aircraft, the right to lay submarine cables and pipelines, and to use any natural resources the nation couldn't use themselves. UNCLOS also states that if the continental shelf juts out beyond the EEZ, a nation has the right to exploit the resources on or within the shelf, within certain defined limits. The debate over the continental shelf is now hotting up in the Arctic, where some parties surrounding the North Pole are arguing over their respective boundaries on the resource-rich continental shelf that lies beneath the Pole.

Following UNCLOS, our tiny subantarctic islands took on a new significance. Our EEZ runs 200 miles from the coast, so thanks to the subantarctic islands New Zealand got over 4 million km^2 of EEZ – one of the top 5 largest in the world – literally overnight.[14] That's 15 times greater than our land mass — which is why over 80% of New Zealand's biodiversity is found in our sea.[15] With this EEZ, we became an oceanic superpower, and a large area of the Southern Ocean really became *Our Far South*.

As a result our EEZ extends all the way to almost 56° South. We also claimed the continental shelf outside our EEZ (remember Zealandia?), extending our territory almost to 60° South, which is the start of the Antarctic Treaty area. In fact, the gap between New Zealand's sovereign area and the waters of the Ross Dependency is a piffling 280 km.

On our territory and within our EEZ, New Zealand has the right to apply our law to manage or totally halt the race for resources. This doesn't mean we always manage things well, but at least we have the ability to do so. This is not the case in Antarctica, because our sovereignty over the Ross Dependency is not universally recognised. Questions of sovereignty are at their most interesting in the rarefied air of the frozen continent.

The Real Race for the Pole

The race for resources in Antarctica was a bit more complicated. At the dawn of the twentieth century much of the world had been claimed by one European power or another, and Antarctica was about all that was left. The only question was how a country could justify claiming it, given that at the time satisfying the occupation component of sovereignty was impractical. This is the story of the real race for the pole.

The British Government in particular was eyeing up adding Antarctica to the Empire. Given that New Zealand, Australia, South Africa and the Falkland Islands were already part of the Empire, control over Antarctica would ensure a virtual British monopoly over the ports of the Southern Ocean. This would mean they could control whaling and shipping routes. Claiming Antarctica proved to be a tricky business.

Didn't Shackleton explore purely for guts and glory? Or in Scott's case, didn't he die dragging piles of rock samples behind him for the good of science? Sure, but part of the reason their expeditions were funded was politics: as part of a bid for British control over the White Continent. The first human occupation of the continent — that is, the first sleepover ashore rather than aboard ship — was by an expedition led by the Norwegian Carsten Borchgrevink, but funded by the British Government. Britain based its Antarctic claims on exploration and discovery. They also used the proximity of its colonies (like New Zealand) to Antarctica to bolster their case. They started by claiming the area south of the Falkland Islands in 1908.

When whalers moved into the Ross Sea in the early 1920s, this finally spurred Britain into declaring full-scale sovereignty over the Ross Dependency on behalf of New Zealand. New Zealand extracted a reasonable income from issuing whaling licenses in the Ross Sea between 1923 (when the claim was made) and 1930.

This nice little earner came to a halt with the creation of factory ships, which were able to process a whale while remaining at sea. In these pre-UNCLOS days, the sea was open slather: you no longer needed a license to whale.

This seemed to spark an Antarctic lolly scramble. The French and Norwegians both made claims based on their exploration of the Continent and in 1933 Britain stepped up their own Antarctic land grab by claiming almost half the continent on behalf of Australia. During World War II, Argentina and Chile both claimed parts of the Antarctic Peninsula, their claims overlapping with previous British claims as well as with each other's. Even Hitler's Third Reich got in on the act in 1939, claiming almost 600,000 square kilometres which it named New Schwabenland, demarcating it with metal darts emblazoned with swastikas and dropped from aircraft. This claim, at least, was extinguished with the defeat of Nazi Germany in 1945.

Claimant nations did everything they could short of permanent occupation to legitimise their claims. They felt that discovery and exploration, if backed up with acts of ownership like licensing whaling and establishing a post office (for who? the penguins?) would do the trick. But this wasn't enough to convince other countries, notably the United States.

The United States was meanwhile starting to get in on the act itself, and to explore Antarctica, too. They used Christchurch and the Ross Dependency as bases for these missions, a relationship that has endured until today. We now view this relationship as an economic blessing, but at the time the New Zealand Government grew increasingly worried. It seemed inevitable that the US would establish a permanent base, and its research stations had a habit of being suggestively named 'Little America'. Remember that in international law, occupation is nine-tenths of sovereignty; there was concern that a claim over the Ross Dependency would inevitably follow.

Through Admiral Richard E. Byrd, the Yanks managed to undertake the first flight over Antarctica in 1929, which allowed them to explore (and therefore possibly claim) a vast territory. In fact, while the US went to great pains not to make any claims or recognise any made by others, they hedged their bets by dropping canisters with a flag and declaration of a United States claim out of the plane onto many remote parts of the frozen continent (including bits already claimed by others). This heraldic carpet-bombing stunt took the farcical nature of international law to new heights.

Post-World War II, the world changed. If the British Empire was drunk with expansionist intentions before the war, they were very much in hangover mode afterwards. Their focus was on holding onto their crumbling Empire, and the United States and Soviet Russia were the new superpowers. The Yanks seemed to signal their intentions toward Antarctica by staging a massive aerial invasion of the frozen continent, code named Operation High Jump, directly after the war. Meanwhile Argentina and Chile started building permanent bases on the comparatively hospitable Peninsula as a way to shore up their claim; a trend that other countries swiftly followed.

Britain could no longer afford to mount Antarctic expeditions with the same frequency. As a small nation, and one much depleted by its own war effort, New Zealand was similarly frugal. After all, we had only taken over the Ross Dependency because Britain had told us to. But three forces soon brought *Our Far South* front and centre in our political consciousness: the fear of a United States claim; the International Geophysical Year (July 1 1957 to December 31 1958); and Sir Edmund Hillary. Having conquered Everest, Hillary was approached by a British chap called Vivian Fuchs to support him on what he proposed to be the first successful overland expedition to cross Antarctica, scheduled to be accomplished during the International

Geophysical Year (IGY). The New Zealand Government had given grudging approval for a small contribution to the expedition, but once Hillary was on board, his profile eventually embarrassed the Government into becoming a full partner in the expedition.

This changed the game. Not only did it lead to the establishment of a permanent New Zealand presence in the Ross Dependency (Scott Base), but it galvanised Antarctica in the hearts and minds of the public. After decades of lobbying by the Antarctic Society, Antarctica finally became an election issue, and it was Hillary's gritty individualism that sealed the deal. Fuchs's expedition planned to cross the continent from the British claim in the Atlantic, past the Pole and out through the Ross Dependency to New Zealand. Hillary's job was to establish supply depots in the Ross Dependency for Fuchs to use. His task completed, a restless Hillary then ignored orders to stay put and set off on converted farm tractors to make a dash for the Pole. While he was ostensibly part of the Fuchs British expedition, to Hillary it had apparently become a race – however friendly – between the Poms and Kiwis. On 4 January 1958, forty-six years after the Amundsen/ Scott race, Hillary reached the Pole, where he found a bunch of Americans lounging around watching a Western in the base that they had recently set up with a series of airlifts. Hillary became the third person to reach the Pole overland after Scott and Amundsen, the first to get there on mechanised transport. New Zealand was jubilant: it was the first time our young nation truly looked toward *Our Far South*.

Of course, none of this necessarily helped convince other nations to recognise our claim over the Ross Dependency. Besides their base at the Pole, the United States had also sited a permanent base in the Ross Dependency, and to further complicate things, in 1956 the Soviets built their first Antarctic base in the area claimed by Australia.

The question now was: should we encourage the United States to claim the unclaimed part of Antarctica (it's still unclaimed today) to lock out the Russians? Or would that put the Ross Dependency at threat from a United States claim? Should we include the Russians to avoid inflaming Cold War tensions? Kiwi diplomats eventually helped to come up with a unique and ground-breaking solution: the Antarctic Treaty.

The Antarctic Treaty – The End of the Race?

The New Zealand Government was in a bit of a bind. Hillary had rekindled Kiwi interest in Antarctica, but safeguarding our interests came with a considerable bill attached. We could barely afford to keep Scott Base going, let alone embark on more expeditions. For our own security, we also wanted to ensure Antarctica would never become a flashpoint for Cold War quarrels like Korea, the Middle East or Central America. The New Zealand public also grew increasingly worried about the possibility of nuclear weapons being tested there, and having the ash blowing onto our back doorstep.

The best way to take the heat out of the race to claim territory and to ensure peace in the region seemed to be through a truly international solution, such as putting Antarctica under a United Nations mandate. So this is what Peter Fraser and Walter Nash originally pushed for. This left us in the rather odd position of arguing to Australia and the United States that the Russians needed to be brought into the fold. If we got everyone playing together nicely in the sandpit, the theory went, peace in our corner of the world would be secured.

And that tactic more or less worked. The cooperative spirit that existed in the name of science during the International

Geophysical Year in 1957 gave rise to a groundswell of support for an international agreement on the management of Antarctica. Over a few years of negotiations, our allies got closer to this way of thinking. While the United Nations didn't end up being involved, when a Treaty was negotiated the Soviets were at the table. In 1959, in Washington D.C., 12 countries found a durable solution to the Antarctic question: the Antarctic Treaty. This achieved our aim of insulating the White Continent from politicking and cold war tensions, and we played a major role in sealing that deal. New Zealand's status as a champion of Antarctica was by now well and truly established.

Let's have a look at a few key features of what was actually agreed in the Antarctic Treaty of 1959.

Freezing of the status quo – all partners agreed to disagree over the question of sovereignty in Antarctica. Countries with claims could continue to believe in them. Those countries that didn't recognise any claims or believed they had a basis of claim of their own (like Russia and the States) could continue to believe that, too. It did nothing to solve the disagreements, but putting all positions into abeyance was truly a genius piece of diplomacy, and a rare, unqualified triumph of international politics. While the Treaty is in operation, no country's actions will count towards future claims, so everyone can relax and stop worrying about the need to mount expeditions or build permanent settlements in the hope of bolstering their claim. The continent is managed through the consensus of Treaty partners, and there are to be no new claims or expansion of existing claims.

Demilitarisation – military personnel and equipment are allowed on Antarctica, but not for military purposes. In other words, they can be used for logistical (supply) operations, but not fighting. Nuclear explosions were explicitly banned (thanks to the Soviets) – but there was room left for nuclear reactors in bases or vessels, so Antarctica is not technically 'nuclear free'.

Importantly, all countries have the right to inspect each other's bases and activities at any time.

Science – the Treaty obliged sharing scientific results, although it is vague on exactly when and how this would happen in practice. Whether it was intended or not is unclear, but the Treaty effectively made science the 'currency for diplomacy' on the continent. In effect, nations outside the inner circle of 12 original Treaty signatories earn their place at the negotiating table by showing a commitment to science on the continent. This has since proved to be important.

Expiry – the main parties to the Treaty are locked in and can't withdraw. However they were entitled to request a review after 30 years (which has now passed). Any changes must be agreed by all.

The box below gives a slightly more detailed summary of the Treaty, for those who are interested.

Quick Summary of the Treaty

Article 1 – Antarctica is to be used for peaceful purposes only. Military activities, such as weapons testing, are strictly prohibited. However, military personnel and equipment can be used for scientific and other peaceful purposes.

Article 2 – Countries shall work together in cooperation for scientific discovery.

Article 3 – Countries shall freely exchange information and personnel, cooperate with the United Nations, and cooperate with other international agencies.

Article 4 – No signatory is required to recognise any territorial claims nor are the claimants obliged to renounce their claims, and no new claims shall be asserted while the Treaty is in force.

Article 5 – This Treaty prohibits nuclear explosions and the use of the Antarctic for the disposal of radioactive wastes.

Article 6 – This Treaty includes all land and ice shelves south of 60 degrees south, and reserves rights to the high seas.

Article 7 – All countries that abide by this Treaty have free access to any area of Antarctica. As such, all countries may inspect any installations, stations, and equipment of other countries, and have free access to aerial photography. However, advance notice must be given of all expeditions and of the use of any military personnel.

Article 8 – Each country has legal jurisdiction over its own observers (and now, by extension, all members of national missions, like scientists).

Article 9 – Regular consultative meetings shall take place between consultative member nations.

Article 10 – Consultative member nations will discourage activities by any country in Antarctica that are contrary to this Treaty.

Article 11 – Disputes are to be settled peacefully by the parties concerned, or the International Court of Justice.

Articles 12, 13, and 14 – These articles deal with upholding, interpreting, and amending this Treaty.

The freezing of the status quo is a neat diplomatic solution indeed, but there are drawbacks. The issue of jurisdiction is left wide open. Whose laws apply to whom and where? Under the Antarctic Treaty, all scientists and observers (people checking out other countries' operations) come under the jurisdiction of their own country. Also, under UNCLOS, anyone on a ship comes under the jurisdiction of the flag of the ship (so on our voyage south, we were actually subject to Russian law! Na zdorovje!).

For everyone else, the law is not quite so clear. None of the signatories can agree whether to apply jurisdiction on the basis of claimed areas or people. Instead, they have an agreement to consult with each other, which as you can imagine strikes trouble on contentious matters of law. In ratifying the Treaty in 1960, New Zealand passed a law applying our criminal law to the Ross Dependency (after we signed the Treaty: doing it beforehand might have upset the other countries), but the Attorney-General decides whether to prosecute a foreign national. Every year the Governor General even appoints 'Officers of the Ross Dependency', who are top dog at Scott Base and who are nominally charged with administering our affairs and law down in the area.

So the Treaty has called a halt to the race to claim Antarctica, but there remains a fundamental disagreement over sovereignty. It simply allows countries to disagree amicably, and to work around the issue. This puts the claimant countries such as New Zealand in an unusual diplomatic position. Most of our diplomatic effort goes into making sure that there are no disputes over Antarctica. However, at the same time, we quietly beaver away to ensure that we maintain our claim over the Ross Dependency. Two-faced? Sort of.

A good example of this tension was the deadline for countries to apply for an extension of their continental shelf (ie, a declaration of the right of countries to exploit those areas of their contiguous continental shelf that lie outside the 200-mile Exclusive Economic Zone). All such claims had to be presented to the United Nations before mid-2010. This deadline very nearly brought the Antarctic Treaty and UNCLOS into a head-on clash. Under UNCLOS, countries had to claim their continental shelf by a certain date or miss out; but the Antarctic Treaty system effectively agreed to disagree over all claims in Antarctica whilst the Treaty is in operation. So who can claim the extended continental shelf off Antarctica? New Zealand and other Antarctic claimant countries faced a catch-22 – if they didn't claim their continental shelf they risked missing the deadline; if they did claim, then they risked raising the disagreements that had essentially been put to one side by the Antarctic Treaty, and undermining the very fabric of the Treaty itself. Different treaty partners found different ways of managing this contradiction. For example, Australia and New Zealand both surveyed the seabed as required to make a claim (see Image #1). Australia submitted their data to the United Nations, and received a few polite rebukes from other Antarctic Treaty players, while asking the UN to not consider their application for the time being. New Zealand took the slightly more diplomatic approach of telling the UN that we'd done the survey but that we wouldn't be submitting our data right now, even if we reserved the right to do so in the future. A triumph of truly diplomatic sophistry, or simply recognition that thwarting the Antarctic Treaty is too high a price to pay for expanding sovereignty over ever more of the seabed? A bit of both, most likely, but that's diplomacy.

In short, the Antarctic Treaty put the issues of sovereignty and jurisdiction on ice. New Zealand and other claimants may argue that their laws apply in their areas of claim, but other countries don't necessarily recognise our authority. Through the Treaty system and national laws, some form of law has been cobbled together for Antarctica. The question is how well are these 'laws' enforced and how will countries resolve any disputes that might arise. To the common person, it may seem like a lawless land. At times it is.

Cold-Blooded Murder?[16]

Fact really can be stranger than fiction. You won't see it on an episode of CSI, but the scientists quietly beavering away at the South Pole could be harbouring a deadly secret.

In May 2000, an Australian astrophysicist named Rodney Marks died mysteriously of methanol poisoning while spending the winter at the South Pole. When the spring thaw came, Dr Marks' body was flown to Christchurch. With the body in New Zealand, the coroner felt obliged to launch an inquiry into the death, which ended up running until 2006. Most of that time was spent trying to get information from the US scientific authorities and the contractor Raytheon Polar Services. United States authorities complied with the investigation up to a point, but there was no way to compel full cooperation under our own law. There were also concerns on their side as to whether this would be seen as New Zealand exercising its sovereignty over the Ross Dependency. There were even esoteric discussions over exactly where the death occurred: given the South Pole is at the apex of most Antarctic claims, a few metres either way makes all the difference as to whose supposed 'patch' the possible 'murder' occurred in!

Dr Marks was known for his drinking, but there was no evidence that he had turned to methanol, even in the bleak 24 hour darkness of the polar winter. New Zealand Police thought it was unlikely that he took the methanol intentionally, but there was no evidence how it might have been administered. He may have been deliberately poisoned, presumably by one of the 49 other scientists at the United States research base at the Pole. Police weren't even given a list of the people on the base, and when they found the information themselves, they had no power to interview most of the people as they lived overseas. Apparently the United States authorities conducted their own inquiry, although the results of that are not known.

In the end, the coroner and New Zealand Police were frustrated with the lack of progress, and the whole issue got popped in the too-hard basket. The coroner recommended some sort of process to sort out these sorts of issues in the future, but of course no signatory to the Treaty wants to give up their sovereignty to another nation, or to a supra-national organisation like the UN.

Peace and Science or a Race to Build Stations?

Under the Treaty, countries with an interest in Antarctica agreed to disagree over sovereignty on the continent. They also agreed that nothing that happened while the Treaty was in operation would impact on territorial claims. This was supposed to take the issue of claims off the table. But as we have seen, it hasn't managed to do this completely: we know existing claimants are still keen to safeguard their claims in case the Treaty ever comes unstuck. So why would other countries completely give up on their own ambitions?

Remember the Treaty System was set up back in the 1950s and 60s, when the Cold War was raging and Sean Connery was James Bond. The Treaty reflected the times, which according to the leading authority on the era (Bond) involved espionage between the US and Commies, martinis (shaken, not stirred) and gadgets embedded in cufflinks. Times have changed. The Treaty may have removed the risk of conflict, but the new kids on the block (Brazil, China, India to name a few) are still looking to flex their newfound influence. It's not about sovereignty any more: it's about status and power.

This battle for status is far more covert than the race for claims. The Antarctic Treaty made science the currency of diplomacy on the continent, so in effect any country wanting to be at the table of Antarctic negotiations (apart from the initial 12 which, thankfully, includes New Zealand) needs a serious scientific programme in place. This focus on science may have created a new race: to build scientific stations.

A lot of incredible science gets done in Antarctica, but there is clearly more to it than the pure pursuit of knowledge. Are these stations really being built so countries can have a seat at the table of Antarctic Treaty partners? Or could it be for national pride — after all, Antarctic science is a bit of a poor man's space programme? Or are they also securing their long-term interest in claiming a stake in Antarctica (permanent occupation being the best basis for a claim)? Who knows. No country in its right mind is going to 'fess up to an agenda that is not in the spirit of peace, love and science: that would be self-defeating.

Since World War II, the number of stations has risen steadily, with two periods of rapid growth. One in the 1950s (while the rush to make claims was still on) and the other in the 1980s (when there were negotiations over the mining of Antarctica). As can be seen in the image below, there are now 83 operational stations on the Continent, with another one planned for 2014. If we include camps, refuges and depots, there are 114 bases in total in Antarctica and surrounding ice shelves.

Map of Antarctic Stations

Image Courtesy of COMNAP, compiled in 2009 by RIA Mobile GIS and Latitude Technologies

There is a strong argument to be made that these bases are not just about the science. Take Scott Base as an example.

Keeping the lights on at Scott Base is our number one Antarctic priority, and a major foreign affairs focus. Our whole Antarctic operation cost about $13.5m each year, of which $5.5m is spent to keep Scott Base itself going. Spending on science varies, but in most years between the Ministry of Science and Innovation and Universities we probably spend an additional $7 million on conducting scientific experiments there. So it is difficult to justify the cost of running Scott Base on the science alone. However, in Antarctica our scientific and diplomatic goals reinforce each other, and from this perspective our Antarctic programme looks like a better investment.

There is no way that we would consider amalgamating Scott Base with McMurdo: even though this may make sense cost-wise, it would dent our claim on the Ross Dependency. Instead we do the next best thing. We have pooled logistical resources with the Americans to keep our Antarctic programme alive: they provide ships and landing facilities and we let them use Christchurch as a base for their programme.

The wind turbines we put between McMurdo and Scott Base were part of this resource-sharing exercise. Fuel is a major problem in Antarctica – it costs a lot to get it there, there are risks of leaks and spills, and there are the emissions from the fuel itself. McMurdo base briefly flirted with nuclear in the 1960s, but it didn't work too well in the cold and the aptly named 'Nukey Poo' was removed along with a few thousands tonnes of contaminated soil. McMurdo and Scott have largely been running on diesel until the New Zealand Government asked Meridian energy to install wind turbines down there. After Nukey Poo, the Americans were naturally sceptical about whether it could be done in such tough conditions, but a bit of Kiwi ingenuity (and American grunt) did the trick. The three turbines, producing 330 kW each, are enough to power all of Scott Base and 15% of McMurdo.

This theory that science and diplomacy go hand in hand in Antarctica actually seems to be borne out by the facts. A recent study showed that those countries with the biggest scientific programmes also tended to have the most say over the governance of Antarctica. This study also showed that the original claimant countries, along with the USA and Russia, still hold the greatest sway in Antarctica in terms of science and diplomacy. And when you take into account New Zealand's size, our contribution far exceeds that of any other player.[17]

Probably the best example where the science is clearly not all they are interested in is Argentina. They are clearly over-compensating for something: they have the highest number of bases, 14 in total. As discussed in the box below, this is presumably to shore up their territorial claim which overlaps with that of Chile and that of their traditional enemies: Britain. After all, under international law, a permanent settlement ultimately equals sovereignty.

Argy Bargy

Argentina and Britain have scrapped over the Falklands since their discovery in the 16th century. Known to the Argies as Las Malvinas, they have proven to be a flash point to this day. In 1982, the pot boiled over into the Falklands War.

A very similar fate almost befell the Antarctic Peninsula. Britain used its Falkland Islands claim and exploration of the region to claim the Peninsula and surrounding area in 1908. During World War II, while Britain was tied up with the Battle of Britain, Argentina and Chile added their claims to the same bit of land. The Peninsula is a strategic area, being the closest part of Antarctica to another land mass.

These competing claims have led to considerable tension between Argentina, Chile and Britain. In 1952, this even culminated in Argentinian soldiers firing shots over the heads of British scientists. All countries have also built large numbers of bases in the region, and even kept a military presence (despite the peaceful status of Antarctica). Argentina has some 14 bases and Chile 10 (with plans for more) plus other sundry camps and refuges. Argentina eventually agreed to put its Antarctic programme under civilian control in exchange for hosting the Antarctic Treaty secretariat.

Probably the most extreme example of the South American expansionism towards Antarctica was in 1978, when the South Americans started flying pregnant mothers to Antarctica to give birth there! Having ten Argentine and Chilean nationals born on the White Continent was seen as a way of bolstering their claims against each other and the United Kingdom.

These bases do have an impact on the environment. Including bases that are now abandoned, over 120 have been built on Antarctica. The 83 operational stations house around 4500 people in summer and 1000 people in winter. A station is not just a building either – some are practically a town. The stations need to be self-sufficient, with a port or an airfield to receive supplies. The largest base is the United States' McMurdo station, just across the way from our own Scott Base. It houses 1000 people during summer, falling to 250 in winter (see Image #7).

You are probably thinking 'so what?' 120 bases built on a continent 1.5 times the size of Europe? It is nothing. But it so happens that bases are competing with Mother Nature for Antarctica's prime real estate. Most of this construction takes place on the relatively tiny bit of coastal Antarctica that is ice-free, particularly on the Antarctic Peninsula. This area is about the same size as Auckland's supercity – about 6,000 km².[18] This may sound like a lot of real estate, but it's not when you have to share it with almost all Antarctica's wildlife, from moss to seals and penguins — including five million Adélies.

Earlier station construction and operational practices were pretty questionable. Most rubbish and sewage was dumped in the sea or under the ice. The standard mode of disposing of junk in McMurdo Sound was to stack it all on the sea ice before the end of winter so that when the thaw came along, it all just disappeared! Out of sight may be out of mind, but it's not gone for good. The cold Antarctic environment struggles to break down chemical and even organic waste, so there is a tendency for the waste to build up in some areas. At one point, Winter Quarters Bay in McMurdo Sound was considered the most polluted harbour in the world, and around 1-10 million cubic metres of soil are estimated still to be contaminated. Between 1957 and 1973, New Zealand and the United States shared a base at Cape Hallett. When this was first built, we had to bulldoze the Adélie penguins off the area! Thankfully, the site has been remediated and the penguins have come back. Between the 1950s and 1980s, some bases were also built on ice shelves, and they have been allowed to become covered in snow and slowly the ice shelf shunts them towards the sea, before spitting them out into the sea in the middle of an iceberg.

The 1991 Madrid Protocol aimed to rein in these environmental side-effects (although as you can imagine, different countries apply it in slightly different ways). It is still possible to discharge sewage, although the likes of Scott Base at least treat it first.

Under the Madrid Protocol, any major environmental impact (like building a new station) is reviewed by all Treaty partners. This has provided a check and balance to stop member countries doing stupid things, but ultimately it is difficult to stop a country doing something they really want to do. For example, we can't stop the Koreans building a base in the Ross Dependency, so instead we put on our best passive-aggressive smile and say 'Welcome to New Zealand. How can we work together?'

India and the Antarctic Specially Managed Areas (ASMAs)

Another protection under the Madrid Protocol is the creation of Antarctic Specially Managed Areas. The whole continent is a natural reserve, but these areas are, well, extra special. These are usually fragile environments that attract interest from scientists and occasionally tourists also. These managed areas usually get zoned and have standards for use, like certain things can only be done in certain places, and there are limits to the number of people that can visit.

How difficult it can be to work together under a consensus Treaty system was thrown into sharp relief with the recent establishment of an Indian base in the Larsemann Hills of East Antarctica. These hills were going through the process of becoming a specially managed area, which included a facilities zone in an attempt to prevent a sprawl of scientific bases in the region. The idea was that if the bases were close together they could share facilities and reduce their impact on the environment. Trouble is, India already had plans for a base in the region, and the spot where they wanted to put it was outside the facilities area.

India wanted to situate their base at the point where they think the Indian subcontinent was once joined to Antarctica some 130 million years ago. They are looking for signs of the Mahanadi River, which they think flowed between the two continents — another obvious example of how bases are being built to provide evidence for possible future claims to Antarctica.

The other countries in the area objected to the additional environmental impacts the Indian base would bring, and asked India to build within the facilities zone. India countered with claims that other countries' objections were also politically motivated: *"the area is very rich in hydrocarbons and those who are already in that area do not want others to come there."*[19] Australia even offered to host the Indian scientists at their base in their area, but this wasn't deemed practical by the Indians. Sadly, few countries share facilities on the ice.

In the end, because of the need for consensus, India's agreement was needed for the Specially Managed Area to go ahead. India was able to negotiate an exception in the Specially Managed Area for their base. The 'Bharti' base, India's third on the Continent, opened in March 2012.

The Antarctic Treaty may have frozen the issues of sovereignty in time, but it couldn't freeze the world order. Times they are a-changing, and to reflect their growing stature, the emerging powers are making huge investments in Antarctica which probably have deeper motivations than the sheer joy of scientific discovery. There are multiple agendas in play: perhaps countries just want a seat at the Antarctic Treaty table; perhaps they want prestige; or perhaps they are eyeing up a slice of the White Continent against the day the Treaty fails and Antarctica gets carved up properly. In short, you can think of the Treaty as a bit like the referee in a rugby game. It may have stopped the fights and got everyone playing reasonably nicely, but it can't stop all the macho posturing.

A good recent example is China's Kunlun station. This has been criticised as being designed for prestige rather than scientific value – at 4 km above sea level, Dome A it is the highest base in Antarctica. Certainly it's the best site for monitoring and managing the flight paths of China's satellites. It's purported to boast a 'Welcome to China' sign, despite sitting smack bang in the middle of the Australian Antarctic claim. The choice of site is apparently to allow studying of astrophysics and the impact of 'extreme weather conditions'[20], but are there other motives?

Another great example of the shifting world order is being played out over icebreakers – the ships with specially designed hulls that can break through the sea ice and that are used to clear a channel to the Continent at the beginning of each summer. The United States has underinvested in icebreakers, and is now dependent on Russia to keep McMurdo Sound free of ice. By association, we too are now dependent on the Russians. Meanwhile the Russians, Koreans and Chinese are all investing in state-of-the-art icebreakers to operate in Antarctica.

Nor did the Treaty completely halt the race for resources; these issues were overlooked during the Treaty process. Without any clear law to resolve disputes, additional agreements were needed to cover issues like mining, fishing and tourism (whaling was already covered by the International Whaling Commission).

For each of these issues, New Zealand and other Treaty partners faced the choice between taking a hard line and opting for total protection of Antarctica, or managing and mitigating the worst aspects of commercial exploitation. Mining, whaling, fishing and tourism are all managed quite differently across this spectrum, and some burning issues are far from resolved. Let's have a look at some of these.

Mining for Answers

'I would not give a nickel for all the resources of Antarctica.'
– Geologist Dr Laurence Gould, 1960

The original Treaty did not address the idea of using Antarctica's resources. At the time any resources that might be hidden under the ice seemed impossible to access. However by the 1970s extraction techniques were improving, it no longer seemed like science fiction that someone could mine Antarctica.

The prospect of mining raised concern about the environmental impacts. It also worried countries with territorial claims, as other countries' mining might be taking 'their' resources and may even lead to their making a claim themselves. Treaty partners realised that they could brush the issue under the carpet no longer. Mining at the very least had to be regulated and managed. In 1976 they started by putting a temporary ban on exploration for and exploitation of minerals and in 1981 began negotiating a more permanent fix.

It took seven years to negotiate the Convention on the Regulation of Antarctic Mineral Resource Activities (CRAMRA), which was finally agreed in Wellington in 1988. It established a body to oversee and regulate mining, protect claimant country rights, monitor the environmental impacts and decide where mining was acceptable and what areas should be protected.

The Convention was never ratified, and the actions of countries during negotiations around mining are another reason to be sceptical about the purity of their 'scientific programmes'. In the decade during which negotiations over mining were taking place (1981-1991), a lot of countries suddenly became very interested in Antarctica. The number of signatories to the Treaty rose rapidly from 23 to 40, and the number of stations in Antarctica went from 42 to 70 (see the graph below). Did someone say 'gold-rush'? Still, more countries signing up to the Treaty was probably a good thing, rather than having them snipe at the régime from outside.

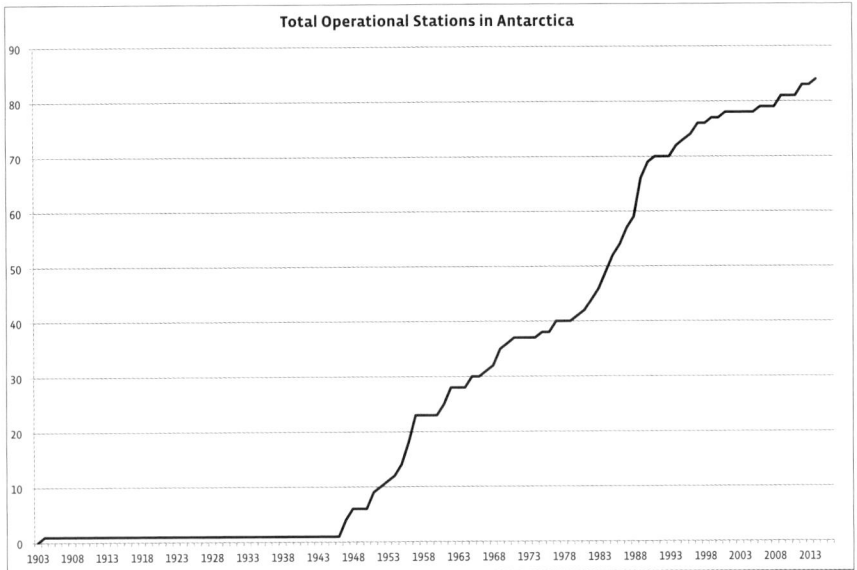

Total Operational Stations in Antarctica

Source: COMNAP

But the Convention was cut up faster than a baby seal in killer whale-infested waters. Environmentalists went nutty. Although mining was to be tightly regulated to minimise the environmental impact, they argued that there was no need for any mining of Antarctica at all, full stop. Other, more moderate critics pointed out that the history of international management of resources was chequered with failure. Either way, the Convention wasn't ratified. Ironically, given how much they mine their own homeland, it was the Aussies under Bob Hawke who dug their toes in and really sank the idea of regulated mining. Some have questioned whether their motives were truly green, whether they were just concerned about protecting their claim, or whether Hawke's stance on Antarctic minerals was a sop to the Green movement that was causing trouble for him in Tasmania.

This still left a vacuum in the management of mining on Antarctica, and just about caused the whole Antarctic Treaty system to fall apart. New Zealand got in behind Australia on the call for a ban on mining, but the US, UK and Japan all rejected the idea. Eventually they bowed to public pressure, and in 1991 the so-called Madrid Protocol on Environmental Protection to the Antarctic Treaty was signed. This banned mining from the continent completely. So in a matter of a few years, the international consensus flipped from favouring sustainable management of mining to a complete ban. As we have seen above, the Madrid Protocol also regulates *any* activity with an environmental impact on the landscape, and allows for setting up Specially Managed Areas.

This Protocol becomes liable to review in 2048 (fifty years after it entered into force). From that date, any party can call a conference to discuss changes to the Protocol, or the scrapping of it altogether. This would still be difficult to do: two-thirds of the original signatories must agree to a review, plus all the countries who were involved in the creation of the Protocol must agree to any changes (so New Zealand has a veto). And it would still need to be replaced by some other means of managing the mining, like CRAMRA. There is one loophole: once all diplomatic avenues are exhausted, countries can withdraw from the Protocol. This looming loophole has people worried – hence the creation of celebrated Polar pedestrian Robert Swan's somewhat ill-named 2041 campaign (Swan thought that the Protocol 'expired' in 2041, but actually it becomes open to review in 2048).

His sloppy research aside, Swan has a point. We certainly need to keep an eye on this issue. Apart from the blip during the global financial crisis, oil, gas and mineral prices have been on a steady rise for some time. Demand for these basic building blocks of industry from the new BRIC economic powerhouses (BRIC stands for Brazil, Russia, India and China) seems inexhaustible. This is spawning a whole new goldrush as companies are prospecting for resources in ever more inhospitable terrain and even deeper sea. As the result of new technology and higher prices, it is possible and profitable to mine in more and more remote places.

The Australian Lowy Institute have noted Russia and China's increased interest in the resource potential of Antarctica, and warned that this could pose a threat to the Madrid Protocol in 2048. The Lowy Institute sees this and other developments as a sign that Australia's claim to almost half of Antarctica is coming under threat, and that pressure to mine and drill Antarctica will in the future be the greatest risk to the Treaty.[21] New players are trying to assert their presence in Antarctica, and nervous claimant countries could well respond by stepping up their presence also.

But ultimately, there is a second line of defence against mining in Antarctica: economics. The remoteness and conditions would make mining and shipping difficult, and most of the continent is (currently) under a few kilometres of moving ice which would make for a nightmare for anyone trying to drill there.

Many of the claims that Antarctica or other parts of *Our Far South* (like the Great Southern Basin) are 'mineral rich' arise from a form of hype known as 'El Dorado syndrome'. This is the human habit of overestimating the wealth of unexplored lands. This practice dates back to the conquistadors, who thought that the fabled city of gold always lay over the next horizon. It was a great motivation for exploring, but sooner or later you hauled over the last horizon, still with no sign of El Dorado. Around the world, this quixotic quest is carried on today by exploration companies and governments, each of whom have an incentive to overestimate the mineral potential of any given place.

It might seem reasonable to assume that a continent the size of Antarctica would hold mineral wealth, particularly since it once butted up alongside mineral-rich bits of earth such as Aussie and South Africa. But most of the claims about Antarctica's fabulous mineral wealth are based on almost zero knowledge. In most places, it is too difficult even to do the exploratory work required. The same applies to the possibility of hydrocarbons in the ocean. Some have claimed that places like the Ross Sea could be part of a massive Antarctic oil reserve.[22] If there are oil reserves, then none of the tell-tale signs (such as oil seeps) have been found.

The only confirmed, sizeable mineral deposit on the whole continent is the coal in the Transantarctic mountains. Trouble is, the coal is of such low quality that even if it were in an accessible location such as Australia, say, it would be unlikely to be mined. Humans would have to be pretty desperate for fossil fuels if we were to go all the way to Antarctica to mine low-grade coal that barely burns.

Of course, we can never say never. The technology has been developed to mine and drill in the Arctic. Rising oil and mineral prices will only increase the incentive to find new technology and sources. However, between the diplomatic barriers imposed by the Madrid Protocol and the economic barriers, mining in the Antarctic seems a distant possibility for now. The biggest risk to the ban on mining might be the El Dorado syndrome itself. If there is enough speculation about Antarctica's mineral prospects, then more countries might ask for a review come 2048.

Some countries are relatively open about the fact that their 'science' in Antarctica lays the groundwork for resource exploitation. In their 20-year strategy the Russians signalled their aim to *'strengthen the economic capacity of Russia through the use of marine biological resources available in the Southern Ocean, and complex investigations of the Antarctic mineral, hydrocarbon and other natural resources'*. To achieve this, they intend to undertake *'geological and geophysical investigations of mineral and hydrocarbon resources on the continent of Antarctica and in surrounding seas'.*[23]

This is reminiscent of the situation facing the Arctic, where the ambient temperature isn't the only thing that is hotting up. The melting ice in the Arctic is allowing access to a range of new resources such as fish, oil and gas, as well as a lucrative new shipping route from the Pacific to the Atlantic. Drilling is beginning in some areas: the Russians are even rumoured to be building the first ever self-contained indoor 'city' in the Arctic Circle, based on the design of the International Space Station, which will house 5,000 residents to man the huge drilling operation[24]. Under the melting Arctic ice cap is the Lomonosov Ridge. The promise of mineral riches has led all the neighbouring states to lay claim to the ridge. The most farcical point of this rift over ownership of the continental shelf came in 2007 when two Russian submarines ventured 4 km below the ice of the North Pole to plant a flag. This mission was intended to boost Russian claims to the seafloor around the pole. The flag was made from rust-proof titanium, just to make sure it lasts a few thousand years. This sort of posturing makes the Antarctic Treaty look positively mature by comparison.

In short, there may or may not be resources in Antarctica. Thanks to the Antarctic Treaty system and the difficult conditions facing miners, most countries seem content to retain the ban on mining for now. But they also want to keep their seat at the Antarctic table, in case this situation ever changes.

The more realistic possibility of striking oil in *Our Far South* is the current prospecting for oil and gas in our own EEZ. At 500,000 km², or one and a half times the size of New Zealand's land mass, the Great South Basin to the south-east of Bluff is New Zealand's largest petroleum basin. So far exploration, rights have been granted for about 18% of this area.

Given the size of the waves in the Southern Ocean, the risks of drilling for oil here seem to be unconscionably high. In fact, in the past, attempts to drill test wells have struggled. Parliament is currently debating standards to govern the environmental impact of oil exploration and extraction in New Zealand waters. Given how delicate the subantarctics are, however, and how risky deep-water drilling in these monstrous seas would be, this protection may not be enough. There is no account taken of the various ocean conditions that occur in New Zealand waters. Perhaps this is what Exxon realised when they pulled out of exploring the Great South Basin shortly after the regrettable Deepwater Horizon incident.[25] The lack of spatial zoning is a clear oversight given the Government's increased focus on extracting minerals and hydrocarbons from the ocean.

Bioprospecting

The ban on mining has prevented large-scale extraction of mineral resources in the Antarctic, and all the environmental damage that goes with it. But another type of resource has become the subject of the latest race: the harvest of genetic material.

This is known as *bio-prospecting*, the practice of searching for valuable substances in biological (living) material. Confused? No wonder. It works a bit like this. With a permit, a scientist can take a sample from Antarctica – a penguin's feather, the antifreeze in a fish's blood or a marine sponge growing on the seafloor. They take that sample back to their country and examine it. They look for substances with useful properties, anything from a cure for cancer to an ingredient for a skin cream. They then patent these substances so that they can benefit from any future use of them.

Environmentally speaking, this practice doesn't cause any more damage than most other scientific research. In fact, everything that happens in Antarctica is no different from other scientific research. It is what happens with the living material when it is taken back to the home nation that raises all sorts of issues. Who owns Mother Nature? Some think it is abhorrent that anyone should.

On the other hand there needs to be some incentive to invest in finding and developing new chemicals that may benefit mankind, even if they come from other organisms. There would be no commercial value in developing a new chemical if everyone were free to use it. A common misconception is that the animals and plants themselves can be patented; this is incorrect. An inventive step is required before any new substances can be patented and used. ZyGEM is an example of a Kiwi company that uses substances from the tiny bugs found on Mount Erebus for a variety of uses, such as rapid DNA testing.

In the case of New Zealand's native flora and fauna – including that in the subantarctics, the status of bio-prospecting is currently subject to a Treaty of Waitangi claim — the infamous Wai 262. The outcome of this process will determine how New Zealand manages this potentially lucrative industry, but it would not apply to Antarctica.

The even more tortuous process of international negotiations on how to manage this issue more broadly is underway. Developing nations such as Brazil are sick of rich Yankee scientists coming to their jungle, sampling strange plants and patenting the latest wonder chemical which they can flog in beauty products, medicines etc. — all without Brazil getting a cent in royalties. Developing nations are arguing for an equitable sharing of the gains from the natural resources in their countries, which is probably fair enough.

Until this issue is resolved, it can't be applied to Antarctica. In the meantime, the Treaty mandates sharing the results of any science conducted on the continent, and the Environmental Protocol manages the environmental impacts of experiments. But once countries take samples back to their own territory, they can do what they like. As a result, companies sponsor science in Antarctica and benefit from the results. Is this right? Surely any discoveries should come with a royalty attached, to help fund Antarctic environmental protection?

Eco-Tourism: An Oxymoron?

Antarctic tourism started in the 1950s and 60s when the Argentines and Chileans started organising short trips across the Drake Passage to the Antarctic Peninsula. Until the 1990s, tourism remained an élite trip, with only around 1000 lucky visitors making it to the ice every year. Since then, numbers have boomed, reaching a peak of 33,000 in the heady days of 2007. The world's economic woes and new regulations banning the use of heavy fuel oil have dampened the numbers of people with the spare cash to splash out on a trip to the Last Continent and forced some operators out of business respectively. The Antarctic Peninsula, in particular, has remained the focal point for visitors, with some spots receiving over 10,000 visitors every year. The Ross Sea is much harder to get to, so visitor numbers remain in the hundreds. Around a third of visitors to Antarctica are from the United States.

The recent growth area in Antarctic tourism is now from 'cruise ships'. These floating monoliths offer passengers a cheaper way to see the ice up close. While most don't allow passengers to disembark and explore, this experience is proving a hit with the ageing population, and cruise ship visitor numbers hit 15,000 in the summer of 2009/10.

All in all, when you add in crew and guides, some 75,000 people visited Antarctica in the summer of 2007/08. Not all of them set foot on the ground, but this still gives a tremendous scope for disruption and destruction of landscape and wildlife. While no one has been killed as yet, and no one goes around clubbing seals or penguins these days, this still raises two major issues: environmental impact and the safety of the tourists themselves.

Environmental Impact

Like scientists, tourists tend to congregate in the small number of ice-free coastal areas on the continent. As we saw above, this risks damaging the delicate and rare ecosystems that exist in these locations. It's hard to know whether tourism impacts on the wildlife. There is some evidence that tourist visits create stress in some animals, causing them not to breed or even abandon their eggs and chicks. For example, the Gentoo penguin seems to breed less when there are humans are around, whereas the Adélie penguin may even respond positively to the extra attention.[26] Kinky little penguins!

Passenger ships may not make landings, but they are far from harmless. Two tourist ships have sunk in Antarctic waters, endangering human lives as well as threatening catastrophic environmental damage with the resulting fuel spill. The sinking of the *Bahia Paraiso* in 1989 spilled 600,000 litres of diesel fuel into the ocean – roughly double the amount of fuel spilled by the recent *Rena* disaster in the Bay of Plenty. Lucky it was diesel, which biodegrades reasonably well, rather than the kind of heavy fuel oil the *Rena* was carrying. Thanks to lobbying by New Zealand and Norway, the worst kind of heavy fuels have been banned from Antarctic waters since August 2011. This will have the happy side effect of keeping away the really massive cruise ships: this may have contributed to the aforementioned recent drop-off in tourist numbers.

Because of the financial crisis and the changes in fuel oil regulations, we are generally seeing newer ships visiting the Antarctic, and the older ones are being retired. However, newer ships don't necessary equal a reduced risk of accidents, either. As ice retreats in the Arctic and the Antarctic Peninsula, some cruise ship companies are getting bolder about sailing in the region without icebreakers and ice strengthened hulls. Add to this the fact that charting of the area is poor, with very little of the Antarctic area charted to modern standards. That could increase the risk of accidents; even if the fuel oil is less harmful, it still poses environmental risks, as well as a massive rescue headache. Treaty countries have been working on a way of determining who would pay for the clean-up if something went wrong, but it is yet to be implemented.

Over time, the Treaty has developed mechanisms (like the Antarctic Specially Managed Areas, which control visitor numbers) to deal with the environmental risks posed by tourism. Tourist operators also have to apply to Treaty signatory countries for a permit to operate in Antarctica. The level of scrutiny to which an operator is subjected ultimately depends on which country they are dealing with. In New Zealand's case all tourist vessels leaving our ports have Government observers on board to monitor environmental performance. Some countries are not so stringent, and if there are problems, then enforcing any action can also be difficult. Just imagine if a tourist operator is based in one country, charters a ship with a different flag and they apply to a third nation for the tourism permit. You can only imagine how the international negotiators would sort out that mess.

To head off problems that might happen as a result of these loopholes, 90% of the tourist operators have formed the

International Association of Antarctic Tour Operators (IAATO). They aim to minimise the impacts of tourism, and have agreed guidelines on how tourist operators should act at certain sites, such as ensuring that only 100 people visit a certain landing place at a time. They seem to take their role pretty seriously, probably because they know if a tourist operator trashes part of Antarctica, they could ruin the party for everyone. They even do a bit of self-policing through measures such as tracking tourist vessels. Anyone who breaks the guidelines can have an observer placed aboard or their accreditation removed.

Safety

As with most attempts at self-regulation, IAATO has been criticised by some as a tokenistic industry response to prevent regulation. The façade that they have everything in hand got a little dented by the sinking of the cruise ship *Explorer* in 2007. The vessel had a history of failing inspections and when their lifeboats were put to the real test they came up short: they floated OK, but the engine on only one of them would start. The crew launched the expedition Zodiacs and used these to tow the liferafts and the helpless lifeboats. This wouldn't have been possible in bad conditions: only the unusually benign weather saved a potentially large loss of life. This got the international community wondering if self-regulation is really enough. The International Maritime Organisation is now working (slowly) on a new code of standards for polar vessels. However, many ships are under a flag of convenience – the *Explorer* was registered to Liberia. As we will see with fishing, even where there is a wholehearted international agreement, it can be difficult to get all nations comply.

The other safety issue that urgently needs addressing is the increased risk of criminal activity between tourists on Antarctica. This could lead to legal disputes between countries, which could ultimately threaten the Antarctic Treaty. It is not hard to imagine situations that would be a legal nightmare to resolve if the countries involved didn't want to cooperate. One person could harm someone from another country, while travelling as part of a tourist group from a third country, which could have got its permit from a fourth country. Finally this might all happen on the claimed territory of a fifth country and the victim might be flown to a sixth country where the coroner or police might decide to investigate. Imagine the legal bills — not to mention the diplomatic nightmare — of sorting that out. Treaty Partners are still working on it.

Until this issue has been sorted Antarctic Treaty partners are unlikely to approve the further expansion of tourism, such as through the creation of full-time hotels on the continent. Besides, tourist operators have plenty of money tied up in floating hotels – cruise ships. Currently only a few operations operate semi-permanent (summertime) facilities in the Antarctic hinterland. These are packed up and shipped off at the end of each summer season, and there are no plans for a permanent facility. In fact, the only possibility of permanent facilities seems to be coming from some cash-strapped Antarctic Treaty partners who are offering tourists bunks in their scientific bases. Uruguay was left red-faced when US inspectors unearthed this practice; there appeared to be very little science happening at one of their bases!

These environmental and safety problems become particularly acute with self-guided tourism. The Antarctic Peninsula is now warming to the extent where it is a viable destination for yachties from South America to pop across and have a BBQ. This is supposed to come under the Environmental Protocol, but as you can see from the story on the next page, that doesn't always happen. How much damage tourists can do is presumably only restricted by their moral compass and the amount of alcohol they can fit on their boat. When things go wrong, it becomes a true test of the Treaty to see how offenders are dealt with.

Going Berserk in Antarctica

Early in the summer of 2010/11, a Norwegian yacht called *Berserk II* docked in Auckland on its way to Antarctica. Our authorities warned them that under New Zealand law, any ship leaving for Antarctica had to complete an Environmental Assessment before they went or face a fine, but the yachties snubbed this inconvenient bit of paperwork. Later on, they also ignored warnings that bad weather was heading their way, which proved to be a disastrous choice.

When they got to Antarctica, two of the crew hopped on some quad bikes and burned off to the South Pole to try to re-enact Roald Amundsen's feat 100 years before. But before they got there a storm struck, and the yacht's distress beacon was set off, some distance from where she had been moored in Backdoor Bay near Shackleton's 'Nimrod' hut on Ross Island. Five nations have search and rescue responsibilities in Antarctica: Argentina, Australia, Chile, New Zealand and South Africa. This incident happened in the Ross Sea, where New Zealand has jurisdiction for coordinating search and rescue efforts. Although this was all happening on the same day as the Christchurch earthquake, we honoured our responsibilities and did our best to direct the search in the area. As it happened, we had an offshore patrol vessel, the *Wellington*, in the area on sea trials, Sea Shepherd had their vessel, the *Steve Irwin*, there and there was a tourist vessel approaching as well. The *Steve Irwin* was closest, and set to work searching immediately. Despite a comprehensive search, all that was found was an empty life raft and some provisions that would have been in the raft. It appeared that the raft, which was pretty smashed up, had deployed from the *Beserk II* automatically, and had never been occupied. The yacht itself and the three crew aboard were lost. The two quad bikers turned around when they heard the news. They made their way to McMurdo Station and were lucky enough to catch one of the last flights back to Christchurch before the base closed permanently for winter.

The New Zealand Government could have prosecuted them, but it was legally and practically much better to pass them over to the Norwegian authorities to deal with. The leader of the expedition, Jarle Andhøy, was prosecuted in 2012 for failing to provide advance notice of the trip, lack of insurance, and failure to submit an environmental impact statement. He was fined the huge sum of NZ$6,000.

This didn't turn out to be much of a deterrent. In the summer of 2011/12, Andhøy returned for more. He was forced to leave New Zealand quickly when it was discovered he was planning another unauthorised trip to the Ross Sea. His plan was to find the remains of the *Beserk II* and work out why the ship lost its mooring. Elements of the Norwegian press have been lionising him for bucking authority and recapturing the 'Wild Viking' spirit. Perhaps some have got a different perspective on pillage to the rest of us. Andhøy, who has a string of environmental offences to his name, was detained by the Chilean navy but then released without any charges being laid. At the time of printing, no further action has been taken for his second offence against the relevant laws and protocols.

Of course, tourism brings positive benefits as well as risks. As our own experience on the *Our Far South* journey made plain, the tourist experience creates a network of people who are passionate about protecting the region. Nature rarely gets a vote even in the most enlightened of democracies, so having people as advocates is absolutely necessary.

Tourism may also be able to provide a source of more tangible support for the environment. Charging for visitor permits can help cover the cost of monitoring and maintaining the areas visited to minimise any harm caused by tourism. Alternatively, volunteer tourism ('Voluntourism') could make a contribution to the monitoring and maintenance of the local environments. As we will see later, it is exactly this sort of basic monitoring that often gets overlooked by national science budgets. This is already the case in the subantarctic islands, where visitors help cover the costs of track maintenance, quarantine and visitor monitoring.

In contrast to Antarctica's challenges, tourism in the subantarctics is easy to manage. That is simply because it is unequivocally ours. The Department of Conservation manages all New Zealand's islands, and there are strict rules governing who can visit and what they can do. All visitors need a permit (there is a limit on total permit numbers), there must be a government representative present at all times, and there are rules to ensure tourists have a minimum impact on the environment (such as sticking to walkways and staying five metres away from all wildlife). If anything, DOC may be managing all this too conservatively. Perhaps it should be opened up to more Kiwis so that they can appreciate the incredible wildlife in the area.

Marine Resources

The one area in which there definitely was a race for resources in the *Our Far South* region was over whales and seals in the early period of exploration of the Southern Ocean. Their populations took a complete hammering and some have not fully recovered as a result. This was not overlooked, and protection was put in place during the mid-twentieth century in the form of separate treaties on sealing and whaling (the Antarctic Treaty itself didn't cover them). But the Antarctic Treaty left the rights of the high seas intact, so after it was signed there were still questions over what would happen with fish.

The whaling issue is still relevant to *Our Far South* today, so even though it is not part of the Antarctic Treaty system, we'll start with a quick look before moving onto fishing. After all, whaling is an excellent example of the collision of the race for resources, environmentalism, and international politics.

Whaling and the IWC

Whales have been hunted for centuries – since the Middle Ages at least. Ancient whaling methods required a great deal of courage. Whales were harpooned and tethered with ropes and floats to small whaling boats as the crew held on for dear life while the stricken animals dived and breached and thrashed in pain. Either the whale shook off its attackers, or it eventually tired. The boat and her crew would then undertake the daring finale — pulling up alongside the whale to administer the final thrust of a lance into a vital organ to kill the whale, ideally without getting knocked out themselves. Using these primitive — to say nothing of barbaric — methods, whaling was relatively sustainable. There was little chance that stocks would ever be harvested faster than they could replenish themselves (although that didn't hold true for the poor old Right whale, which was called the Right whale because it was slow and floated when it was killed, which made it the right whale to hunt). The bigger, faster whales known as the rorquals — the fin, the sei and the blue — were extremely difficult to catch while everyone was still hunting them with sail and rowboats.

The industrial revolution changed all that. First, it created a big spike in the demand for lighting and lubrication oil. And as it wore on, the technology of whaling improved, too. Steamships got faster, and someone had the bright idea of putting a bomb on the end of a harpoon. Not only could you catch up with a rorqual, now: you could actually kill it, too. Not even the introduction of petroleum refining at the end of the nineteenth century saved the poor old whales. Baleen (the marvellous, flexible stuff that some whales have in their mouths instead of teeth and that they use to filter their tiny food from seawater) was in demand for women's corsetry. And after World War I, baleen whale oil became the best source of the glycerine that was used for explosives. The invention of factory ships in the 1920s meant ships didn't even have to go back to shore to process a whale: they could keep on whaling. These technologies were so successful that whalers had to go further and further afield to find their quarry, eventually ending up in Antarctica.

Thanks to krill, the Southern Ocean held the lion's share of the whale populations, particularly around the south of the Atlantic. In the Southern Ocean, around 2 million whales were killed between 1904-1986.[27] The blue whale was among the hardest hit, with about 350,000 blue whales killed, which adds up to more than 30 million tonnes of whale. Each fully grown blue whale weighs about 180 tonnes, the same as 40 elephants. Most were killed on their feeding grounds in the area close to the Antarctic Peninsula — some 2 million km^2 — which translates to a density of one blue whale killed per 6 km^2.[28]

Prior to World War II, some whale populations were already showing signs of strain. For example, by 1915 the humpback whale population around South Georgia was wiped out: whaling had only started there in 1904. Whales are slow-growing and tend to give birth to relatively few calves over their lifetime. This means their population cannot be 'harvested' at anything like the same rate as fish. The harvesting of Right whales was banned in the 1930s. Things didn't let up post-World War II, either: Japan was hungry and whale flesh was seen as the solution. Britain, which had decided to get out of industrial whaling, managed to pass off its old whaling fleet as reconstruction aid. Half of all meat eaten in post-war Japan was whale.[29]

In recognition of the natural limits of whaling, the International Whaling Convention was signed in 1946 with the aim of sustainably using and conserving whale stocks. Initially there was no suggestion of a complete moratorium: the aim was simply to guarantee future whale harvests. A number of lessons were learned from whaling negotiations, mostly about how not to manage international resources. As with most international negotiations, it took time to agree catch limits. Even then, the IWC proved to be a spectacular failure at managing whale stocks, thanks to difficulties in managing different species and even some downright skulduggery.

Because of overexploitation, whaling of humpback and blue whales was banned in the 1960s, but this was not enough to stem the tide. Whaling in the Southern Ocean was managed on the basis of the total weight of whales caught rather than on a species basis, so humpbacks and blue whales kept getting caught.

Massive illegal whaling campaigns carried out by the Soviet whaling fleet after WWII also ensured that numbers of most whale species continued to slide. The IWC agreements were initially put in place with countries monitoring their own vessels. The fox, in other words, was in charge of the chicken coop, and funnily enough, that régime didn't work; records from Soviet biologists (kept secretly and released after the Cold War) suggest that illegal whaling remained commonplace well after the IWC came along, and even into the 1980s until the moratorium on commercial whaling came into force (in 1986). In fact, captains were encouraged to exceed set quotas: for example, between 1946 and 1986 the Soviets killed 48,000 humpbacks and only reported 2,710. This appeared to continue even when the Soviets carried 'independent' Japanese observers. Were it not for the efforts of a few brave biologists, we would never have known about these activities.[30]

Over time, public opinion on whaling began to change. Whales were no longer needed for their oil and meat, nitroglycerine was out of vogue and women's fashions had gone away from full hooped skirts and stays. Whales became the poster child for environmental groups. Countries that were anti-whaling (as New Zealand had been since giving up the habit in 1968) re-joined the IWC and campaigned to stop all whaling. This eventually led to a moratorium on whaling being passed with the necessary three quarters majority. Japan initially refused to recognise the decision, but was persuaded by the United States who were withholding rights to fish in their waters.

There were sufficient concessions and loopholes within the IWC moratorium to allow some countries to continue whaling. Allowing tribal people who depend on whale meat to survive is fairly reasonable (although the right can still be abused), and in 2008/09 around 350 whales were killed for this purpose. The Japanese continued to whale under the IWC's scientific permit scheme (which allows a certain number of whales to be killed for research purposes and the meat to be sold): since 1986, they have killed around 10,000 whales, mostly minkes, which are the smallest of the rorquals. They must know a heck of a lot about whales by now. To put their sudden, insatiable desire for cetacean information into perspective, Japan had caught 840 whales for scientific purposes over the 30 years prior to the moratorium. Japan also refused to acknowledge the Southern Whale Sanctuary (which they are perfectly within their rights to do), which was established at New Zealand's and Australia's behest in 1994. Norway and Iceland have taken the slightly more upfront approach of simply opting out of the moratorium; both continue to whale within their own seas.

Now more than ever, the IWC is mired in politics. Developing nations are sympathetic to Japan's view of whaling as simply the use of natural resources. They also like to point out that their eating of whale meat is little different to our eating of beef and pork (except that cattle and pigs are farmed and usually killed humanely, whereas even the most advanced refinements of the technology of whaling such as electric harpoons haven't made a whale's death anything other than slow and painful). Through a mix of diplomacy and foreign aid, the Japanese have persuaded countries to join the IWC and bolster their voting bloc — one new member nation paid their registration fee to the IWC meeting at St Kitts and Nevis in cash, which they presented in a brown paper bag bearing a supicous resemblance to the one Japanese delegates had been seen with a little earlier on. Japan is close to commanding a majority in the IWC, and may eventually get the necessary votes to re-start commercial whaling.

Even if this were to happen, most whale populations are still far too low to sustain whaling.

There are only around 1,700 blue whales left in the Southern Ocean, which is thought to be less than 1% of the initial population. Probably only the minke whale could be sustainably harvested at the moment, although how many minkes could be harvested is hard to know even for these whales, because we don't have a reliable population estimate or know how fast they breed. Of course, if and when whale populations recover fully, the Japanese could have a point: by fishing some animals and not others, some scientists argue we risk tipping the ecosystem out of balance.

In the face of the prospect of renewed commercial whaling, New Zealand tried to broker a compromise deal where Japan would be allowed to pursue commercial whaling but with a sinking quota (below what they are taking now in their scientific programme). The rationale was that if whaling was allowed but at a steadily reducing rate, it would eventually become commercially unviable and Japan would be presented with an honourable exit from whaling. With only 1% of Japanese still eating whale meat, this seemed a real possibility; the only things keeping them whaling are probably pride and because it is a way of keeping the political heat off their overfishing of bluefin tuna. In the end, no compromise could be found: naturally, Japan wanted higher whaling levels, while as a result of lobbying from environmental groups, some nations objected to any whaling at all. We were getting it the neck from both sides — for not allowing enough whales to be killed, and for allowing any whales to be killed at all. When you get two opposites banging against each other like that, the chance of a compromise quickly disappears. So we're back to the IWC stalemate and a huge opportunity has been lost.

More recently, Australia has taken the unusual step of taking Japan to the International Court of Justice. Their argument isn't yet public, but it seems that they will be arguing that Japan is actually undertaking commercial whaling in disguise. No, really? The Court will probably have to decide what is a 'reasonable' whale harvest for the purposes of a scientific programme.

Meanwhile, whaling isn't the only or even really the main issue facing whales. As we will see, climate change is altering their food supply. Also in some parts of the world fishing nets (particularly gillnets) are a problem; in 1994 nets killed an estimated 200-300,000 cetaceans (whales and dolphins, particularly the small ones).[31] Ship strikes are a problem — the Japanese taunted New Zealand with a statistic that suggested we were likely to be responsible for more great whale deaths than their scientific whaling programme. General pollution of habitat is just as much a problem for whales as it is for anyone who wants to eat them: whale meat is rich in accumulated nasties like mercury and PCBs, and because they are mammals, so is the milk with which they suckle their young. It makes sense to refocus international negotiations on these broader issues.

So with whaling, New Zealand has taken a stance to lock it up completely in order to protect the resource, yet this hasn't been possible because other countries disagree. Instead, Japan favours sustainable utilisation of the resource. Only time will tell how the Aussie court case pans out: it could either bring whaling to an end or it could legitimise the Japanese position.

Fishing and CCAMLR

The Antarctic Treaty left intact a nation's right to do pretty much whatever it liked on the high seas. You guessed it: this led to a race for resources and several Antarctic fish populations got decimated. Soviet fishers in the 60s and 70s plundered stocks of marbled rock-cod (which took only two years to almost wipe out) and mackerel ice-fish around South Georgia. These stocks have not yet recovered, even though many years have elapsed since the fishery was closed.

When the Soviets turned their attention to krill, alarm bells started ringing. If they managed to wipe out krill, the whole Antarctic ecosystem could well fall over. This led to a ground-breaking agreement to manage the seas around Antarctica, known as the Convention on the Conservation of Antarctic Marine Living Resources (CCAMLR, pronounced 'Camel-R'). This came into force in 1982, making it one of the first international fisheries agreements, and the only one that lists conservation of the environment as one of its aims. Importantly, this did not exclude harvesting, so long as it was carried out in a 'rational' manner.

Because of the focus on conservation, CCAMLR has done a lot of work to understand the Antarctic food chain in order to understand the impact of fishing. This is no small task; the environment is very complex and we are still a long way off understanding it. Other things, like climate change, also impact on the environment at the same time, which makes it difficult to isolate and understand the impact of fishing. CCAMLR mitigates this by taking a precautionary approach to setting fishing limits. Their models try to show the risks of different decisions, and they have a number of indicators that are monitored to forewarn of problems in the ecosystem. There are two major stocks managed by CCAMLR: toothfish and krill. Let's start with krill.

Krill populations fluctuate wildly, so catch levels are set so that on average, krill stocks will be 75% of the original biomass, and they will not fall below 20% of the average original biomass in a bad year. Out of a total population of around 125-750 million tonnes, this means a little under five million tonnes can be taken each year. There is also a temporary restriction of 620,000 tonnes in each sub-area, which has been set until fishers can agree on how to share the catch out between them.

As far as fisheries management systems go, this is a very precautionary approach. CCAMLR are being more cautious than they might otherwise have been with krill, because they recognise its importance to the rest of the food web. As can be seen in the graph below, krill catches in recent years haven't approached any of the limits that have been set by CCAMLR and are considerably below historic catches.

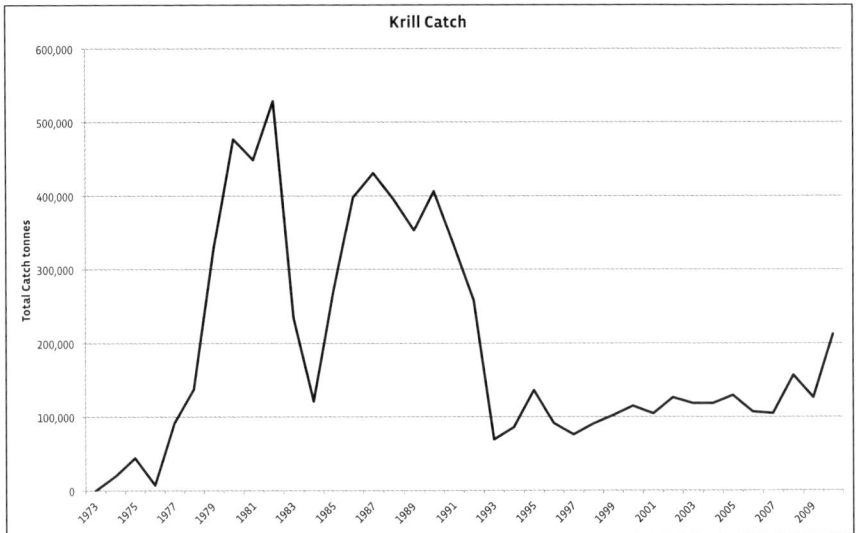

Source: CCAMLR

It looks much like graphs for other fisheries, but for once, the steep decline in this curve doesn't represent the collapse of the fishery: the Soviet Union was krill's biggest customer, and once the Soviet Union disappeared, so did much of the demand for krill. The krill catch peaked way back in 1981 when the Soviets, with their customary focus on quantity over quality, scooped up around half a million tonnes of krill. Since then, the high levels of fluoride in krill shells (fluoride can be poisonous in large quantities) and the fact that krill spoils quickly have made it difficult to turn a profit from the krill fishery. Most krill tends to be used for aquaculture (as feed for other fish), rather than as a food for the masses. For these reasons, krill catches stayed well below levels that would threaten the resource or the wider ecosystem.

New technology has emerged that allows boats to take the shell off the krill automatically, and suck them out of the water continually using giant vacuums. This technology is expensive, and so these krill are used for a special Norwegian Omega-3 oil product. Thanks to this innovation the catch is back up above 200,000 tonnes for the first time since the Ruskies went belly up in the early 90s. If the catch continues to rise this will test the mettle of CCAMLR and their ability to control krill fishing.

What future for krill? It may depend on what we want to use it for. Some experts have questioned whether we should sanction any fishery that is used for aquaculture – precious seafood should be eaten directly by humans or left in the ocean for the whales to have. Aquaculture tends to be wasteful – in addition to the costs of catching and transporting the feed, it actually produces less fish overall than it consumes. In other words, aquaculture raising things like salmon and prawns actually reduces the global stock of wild fish. However, more direct uses for krill are appearing, such as in fancy fish oil supplements destined for the medicine cabinets of the worried well.

On the other hand, some argue that krill is a food for the future. It is a major source of protein and Omega-3 oils, both substances that will be in huge demand as the world's population grows. Eating krill directly could be a more efficient way for humans to meet their nutritional needs. Krill is lower down on the food web and so more abundant. Every time a critter gets eaten, 90% of the mass of the critter getting eaten is lost, so the higher up the food chain you go, the more energy is lost. So it's more efficient to cut out the middle man, so to speak, and go straight to the source. In other words, we could feed a lot more people if we ate small seafood, like krill and sardines, rather than big fish. But this argument only holds if we actually eat the krill ourselves: it falls over if we then just feed it to salmon.

The other major fishery now is the highly lucrative Antarctic toothfish. This has been a controversial fishery – particularly since IUU fishing has decimated populations of its cousin, the Patagonian toothfish.

Again, CCAMLR has regulated this fishery by imposing an annual catch limit, and in the Ross Sea region (as defined here) the catch limit is just under 3,300 tonnes. So far, fishing is believed to have reduced the adult toothfish population to 80% of the unfished level, and this is expected to fall to 50% by 2050, where it will stabilise. CCAMLR allows toothfish to be fished to a lower level than krill because it is considered a predator and not prey. Still, 50% of the unfished level is far higher than the management targets for most fish stocks around the world.

Toothfish is managed using a precautionary approach, designed to preserve the wider ecosystem. Given the importance of the Ross Sea region, this makes total sense. In fact, the New Zealand toothfish fishery is considered so well managed that it has qualified for Marine Stewardship Council certification. This doesn't stop some of those who oppose all fishing in the Ross Sea region from perpetuating many myths about the fishery, which we look at in the box on the next page.

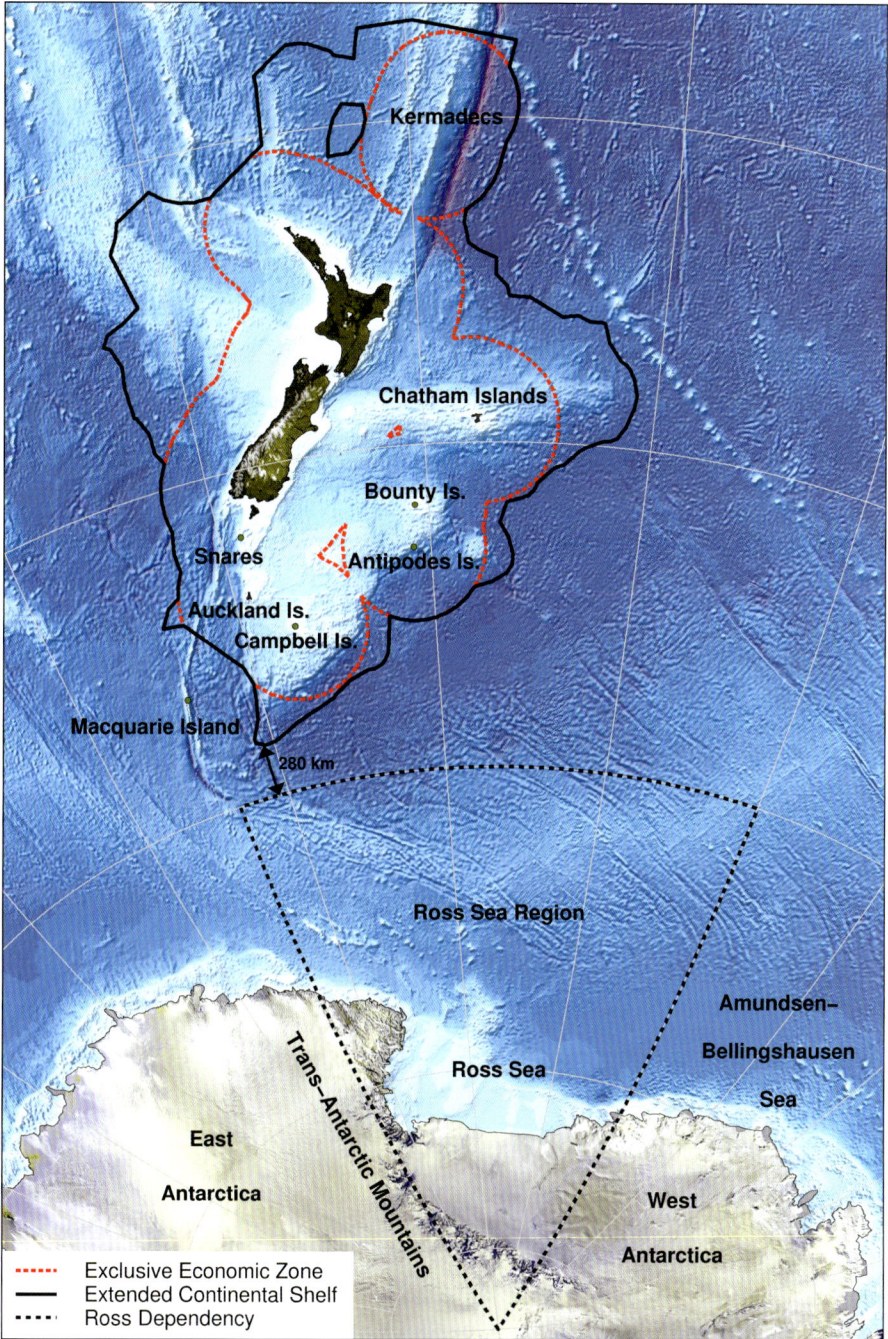

Image #1: Territory of Our Far South

Image Credit: Seafloor data from NOAA, ETOPO1. Satellite imagery from NASA, Blue Marble: Next Generation. Territorial boundaries from LINZ. Plotted by Dan Zwartz using GMT, the Generic Mapping Tools.

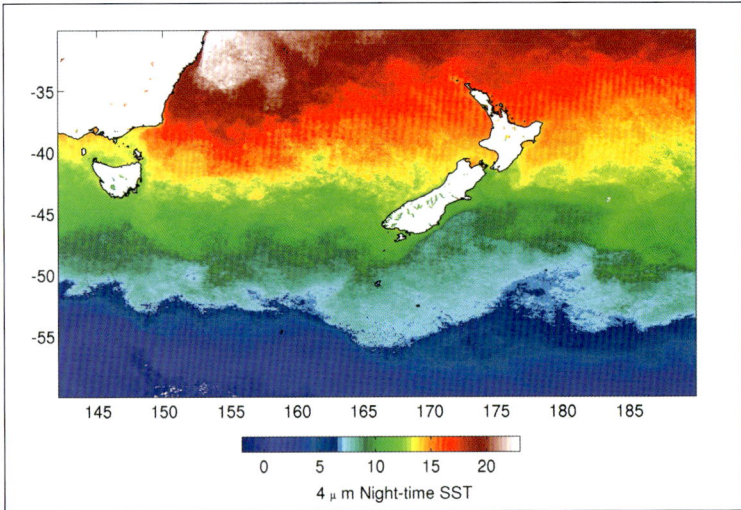

Image #2: Spring Sea Surface Temperatures

Source: NIWA – the Subtropical Front is shown by where the green area meets the blue area.

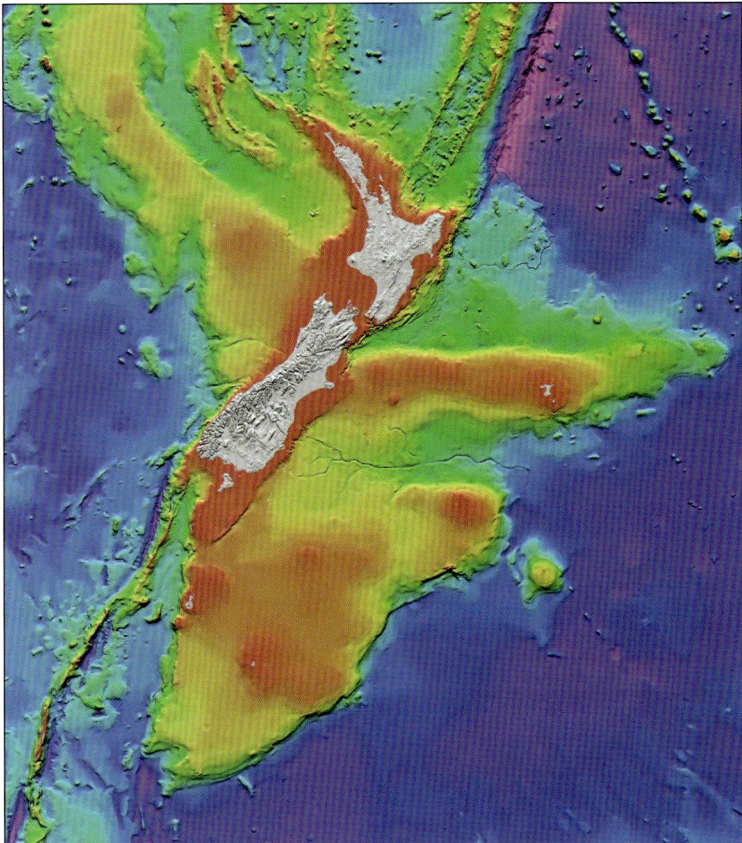

Image #3: Our Great Southern (Underwater) Continent

Source: NIWA – shallow water in red down to deep water in purple

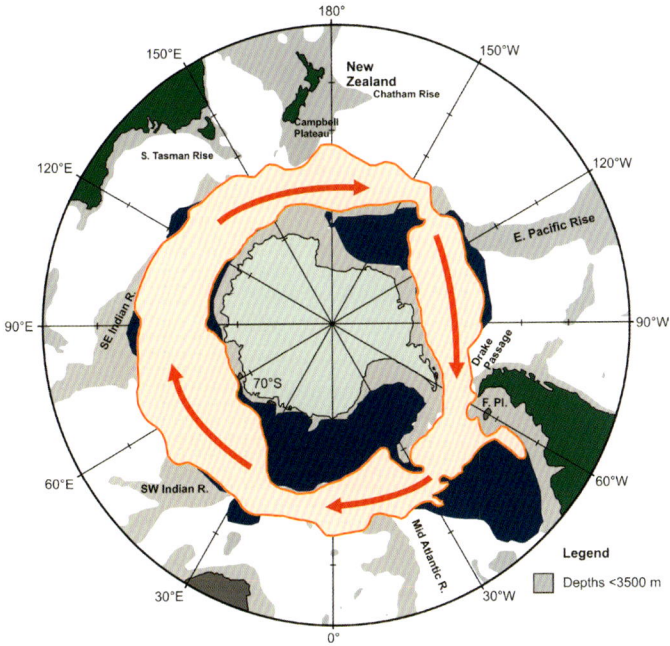

Image #4: The Antarctic Circumpolar Current

Source: Lionel Carter. The ACC area is shaded orange, and the blue denotes where the cold, salty Antarctic bottom water is created.

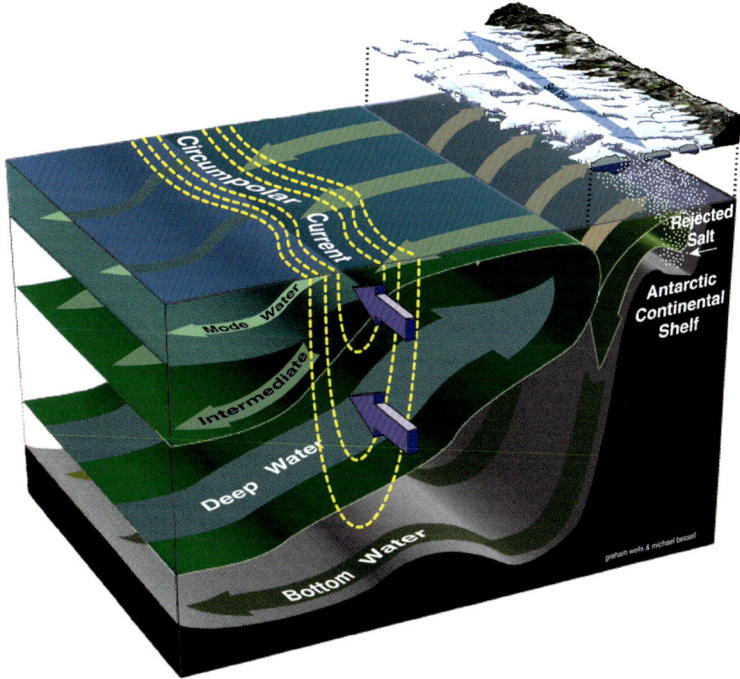

Image #5: Antarctic Currents

Source: ©Steve Rintoul ACE CRC. From: Rintoul, S. R., 2000. Southern Ocean currents and climate. Papers and proceedings of the Royal Society of Tasmania, 133, 41-50.

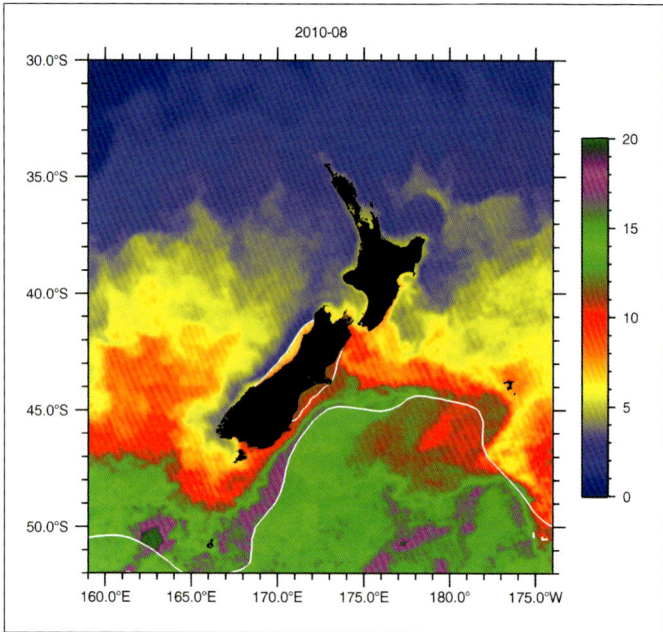

Image #6: Nitrate Levels in the Ocean Surface

Source: NIWA. The nutrient rich waters to the south of New Zealand
(courtesy of the ACC) are shown by the green and purple colours.

Image #7: View of McMurdo Base from Observation Hill

Image Credit: Lance Wiggs.

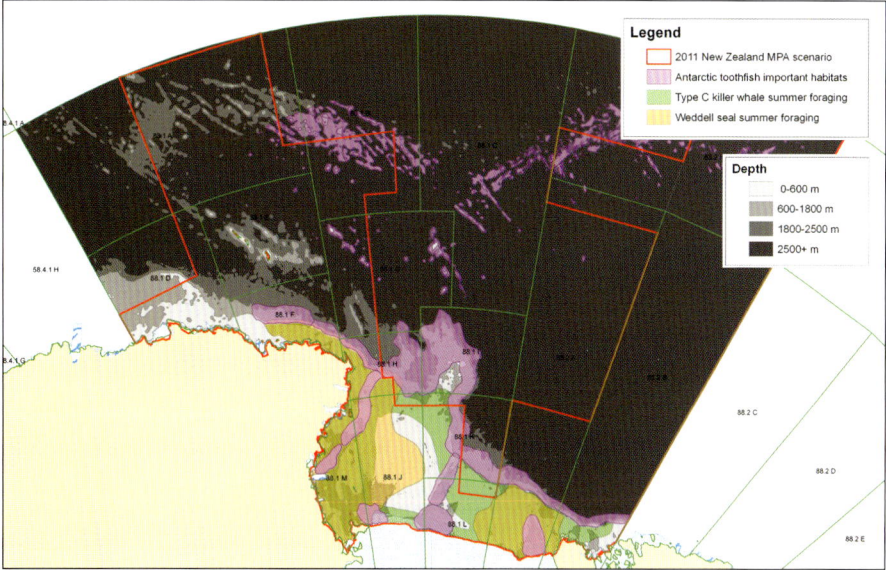

Image #8: New Zealand 2011 draft MPA proposal

Source: Ben Sharp, New Zealand Scientific Committee Representative to CCAMLR. The image shows how the proposed protected area will protect most areas where toothfish predators (Weddell seals and Type C killer whales) hunt.

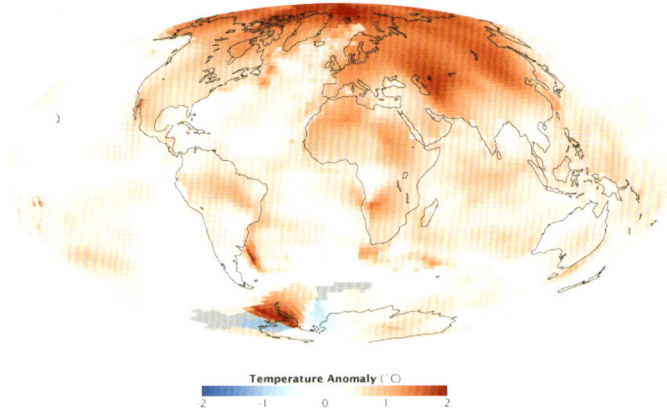

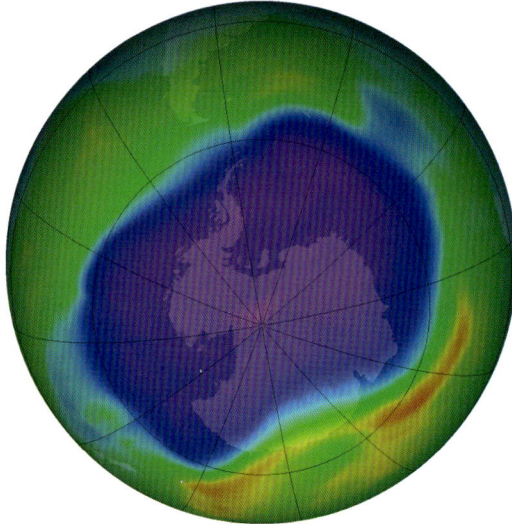

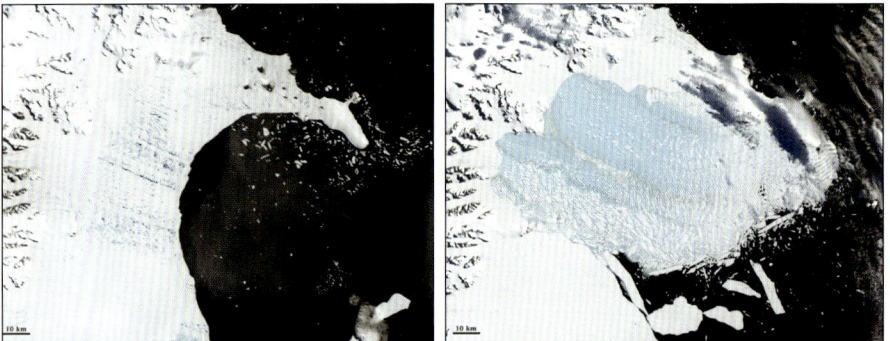

Image #11: Collapse of the Larsen B Ice Shelf

Image Credit: NASA Earth Observatory – The first image was how the shelf was in January 2002, the second shows the sudden collapse of the ice shelf in March, which released 3,250 km² of ice into the ocean. This was the first time the shelf had collapsed in 12,000 years.

Image #12: Enhanced Map of Ocean Photosynthesis
Source: GeoEye Satellite Image – courtesy of NASA/ GeoEye.

Image #13: Emperor Penguins
Image Credit: Su Yin Khoo.

Image #14: Seabed of the Ross Sea
Image Credit: Rod Budd, NIWA.

Image #15: Elephant Seal
Image Credit: Rob Murdoch, NIWA.

Image #16: Antipodean Albatross
Image Credit: Rob Murdoch, NIWA.

Image #17: Rubbish Collected on Macquarie Island
Image Credit: Keith Springer, Tasmania Parks and Wildlife Service.

Myth-busting Toothfish Claims

Antarctic toothfish stocks are in trouble. The scientists maintain that this is just wrong. By far the greater proportion of the evidence suggests that the adult Antarctic toothfish population is at 80% of their original, un-fished levels, on their way down to 50% as per the rules of the fishery. Even this 50% level would make toothfish one of the most cautiously managed fisheries in the world; by contrast, most NZ stocks are managed between 20-35%. If toothfish were managed in the same way, it would probably be fished down to around 24% of the adult population.

Of course, the approach used in New Zealand waters has been criticised as being too risky. It doesn't consider the wider environment and doesn't allow for the fact that in nature, populations are volatile. In an effort to be more cautious, fisheries managers are now targeting 30-40% for New Zealand fish stocks[32], though it remains to be seen if they ever get there. Toothfish aren't too different to many NZ fish stocks, so either way you look at it, 50% seems very precautionary.

These claims that toothfish are in trouble have come from a few American and Kiwi scientists (none of them fish stock specialists) fishing with a line through the ice in McMurdo Sound. The data their claims are based on were found by CCAMLR to contain inconsistencies and problems – some of the numbers didn't add up, and their fishing site changed halfway through the sample. As this book goes to print their data has been re-submitted to CCAMLR, so watch this space.

Regardless, even if their claims hold true we can't take one data set too seriously. Data on toothfish are collected from across the entire Ross Sea, and this shows that the population is holding up well. Anything could have happened in that location to reduce the toothfish there – McMurdo Sound was blocked by a major iceberg for many years, and it is also one of the most polluted places in Antarctica. Using one data set to determine the fate of toothfish is the equivalent of assessing New Zealand's fish stocks by fishing off Queens Wharf and only snagging gumboots. Some of the claims made by penguin scientist David Ainley are an outright slur on the work of NIWA scientists, who are the experts in this area and maintain the highest professional standards and credibility. It reminds us of the climate change issue, where loopy talk-show hosts give credibility to deniers who have never published relevant, peer-reviewed material.

Toothfish are long-lived and slow-growing and therefore more vulnerable than other fish. It depends what you call slow-growing and long-lived, but toothfish are nothing like our own orange roughy, which some like to compare them to. Lessons were learned in the orange roughy catastrophe, and the precautionary management of toothfish is believed to more than compensate for its age and growth rate, which is why the fishery passed the Marine Stewardship Council certification with flying colours. The Kiwi Ross Sea toothfish fishery scored 91 out of a possible 100, compared to our own hoki which scraped through with a score of 81 (the pass score for certification is 80). It would be great if New Zealand fisheries were managed with as much precaution as toothfish. In terms of overall age and time to maturity, toothfish are about on a par with our own hapuka: should we completely stop fishing those too?

We know almost nothing about toothfish. How about we ask an expert? NIWA toothfish expert Dr Stu Hanchet says 'this claim is extremely frustrating and quite misleading. Over 15 peer-reviewed scientific papers have been published in reputable journals and 150 reports submitted to CCAMLR on aspects of its life history including age and growth, length and age at sexual maturity, distribution, diet, condition, life history, trophic status and stock structure over the past decade. A mark-recapture experiment has released over 25,000 tagged fish and recaptured over 1000 fish since 2002. These data have formed the basis for assessment of Antarctic toothfish abundance in the Ross Sea since 2005. In fact, we know more about Antarctic toothfish and have better monitoring methods for the fishery, than we do for the majority of fish stocks around New Zealand.' Thanks Stu. This information is largely thanks to Kiwi fishers working with Kiwi scientists at their own expense to ensure the fishery is sustainable. Our fishing industry is by no means perfect, but the toothfish fishery really is an example of them at their best.

CCAMLR uses many measures to ensure that the total catch stays within the agreed limits. First, they license the vessels that take part in the toothfish fishery. Four of the 15 vessels licensed to fish toothfish in the Ross Sea are from New Zealand; the others are from Argentina, Korea, Russia Spain, UK and Uruguay. Each vessel has to carry two observers. While there is a limit on the total catch, unlike New Zealand's quota system, there is no limit to the amount each vessel can catch. Instead, the vessels compete with each other to catch toothfish until the total limit is reached and the fishery closes. This is known as an 'Olympic' fishery – it's a race to fish every season.

So well managed or not, there is still an element of a race for resources about the toothfish business, and this Olympic fishery does cause some problems. Countries are eager to send vessels down to secure a share of the catch, and some of the vessels they send aren't fit for the conditions. This is probably the cause of the incidents we have seen with fishing boats in the Ross Sea catching fire and getting stuck in the ice (three boats and 27 lives have been lost in the last three years). What is worse is that at the moment fishing vessels are exempted from the agreement on liability for clean-up if there is an accident. The problem could largely be fixed by doing away with the Olympic fishery and giving each country a fixed quota to fish.

So the toothfish fishery, the Kiwi portion at least, seems to be very well managed. Through the use of longlines, the environmental damage is minimal compared to trawling. Harm to other species is also minimised – only one seabird has been caught in the history of the Ross Sea fishery, and there are strict rules in place dealing with the way fishers must behave if they are taking too much in the way of bycatch.[33] Best of all, unlike most managed fisheries in the world, the CCAMLR toothfish fishery has 100% observer coverage, including by international observers. That our fishers have done their best to keep their noses clean in the Ross Sea is recognised by the recent Marine Stewardship Council certification of toothfish caught by New Zealand and the UK in the Ross Sea.

Illegal, Unreported and Unregulated Fishing

As we saw in our book *Hook, Line and Blinkers,* there is a global race for resources over fish, so most fisheries end up with too many fishers chasing too few fish. This excess capacity is one of the key reasons why illegal, unreported and unregulated (IUU) fishing occurs. There are plenty of people who are perfectly willing to take the risk and break the law, particularly if the chances of getting caught are so slim. Some of them are desperate, hungry and highly indebted fishers; others are powerful corporations and organised crime syndicates.

IUU fishing is potentially an enormous threat, particularly in a vast and distant ocean such as that which surrounds Antarctica.

This is partly because monitoring is so difficult, but also partly because of international law. Reining in fishers who aren't playing ball is up to the government that vessel is registered with (which you can tell by the flag they fly). As always, some governments are better at doing this than others.

Due to its nature, it is pretty difficult to know the true extent of IUU fishing. However, there are clear examples of the devastating impact it can have. For example in 2009, Australian authorities discovered an illegal gill net, 80 miles long, filled with 31 tons of dead toothfish.

IUU fishing in Antarctica has been a lucrative business, until recently when great strides have been made to get on top of it. Under CCAMLR, international agreements have been designed that aim to track all toothfish caught and ensure they come from a licensed supplier. New Zealand, working as part of CCAMLR, also works to track fishing vessels and documents all toothfish catch at ports. New Zealand also conducts aerial and naval patrols of the Ross Sea on behalf of CCAMLR and inspects toothfish vessels while in port.

Now CCAMLR is targeting countries that allow IUU fishing vessels to use their flags or ports, and since 2004 has been putting together a list of IUU vessels. Even the EU is promising to get tough with its members who aid and abet IUU fishing. They have also asked all nations to ensure that the fish they import are not IUU, which is pushing countries to trace their fish supplies better.

The signs are positive. However, as we see from the example on the next page, doing something about IUU fishing is not always so straightforward.

Netting the Poachers

There are a plenty of examples of just how hard it is to manage illegal fishing. In 2003, an unarmed Australian customs vessel chased a Uruguayan-flagged toothfish poacher 6,500 km across the Southern Ocean before it was intercepted and boarded by a Falklands-based British naval ship. The poacher got off, as the Aussies couldn't prove that the fish were caught in their waters. Unfortunately, this epic journey is one of the simpler cases, because it was so clear-cut. Usually the chase is more metaphorical, and everyone tends to get trapped in the nets of international law. A more complicated example involves the *Paloma V*.

Ironically, the toothfish poacher in the chase scene above and the *Paloma V* allegedly share the same owner. They are both thought to be part of the shadowy 'Vidal Armadores' fleet – a Spanish company involved in an illegal fishing ring linked to more than 40 instances of illegal fishing. That's right: when it comes to big organised crime, the Italians have the mafia, the Colombians have the drug lords, and the Spanish have illegal fishers. You can imagine them gathering in darkened rooms and smoking… mackerel. Over the years this fleet is thought to have been fined about US$5m for its transgressions, yet it has also allegedly pocketed around US$12m in subsidies from the Spanish Government.[34] ¡Muchas gracias!

In 2008, the *Paloma V* had been fishing legally for toothfish under the Namibian flag when it docked in Auckland and was subjected to a routine inspection by our Ministry of Fisheries. They asked the crew where the catch records were and were referred to the computer. The Ministry officials cloned the entire hard drive and found evidence of *Paloma V* working with illegal vessels – messages to blacklisted ships, a radio network with the illegal fleet and even photos and records of them offloading supplies (such as fuel and, just as important to Spanish pirate captains, cases of wine) to illegal fishing ships.

Thanks to the work of CCAMLR, blacklisted vessels operating around Antarctica struggle to get supplies or to unload their catch because they can't use ports in the region. However, to get around this, they work with legal fishing ships, offloading fish so it looks like the legal boat caught it and in exchange, taking on supplies such as fuel, bait and food. Proving this practice is happening is pretty difficult given the remote location.

After a bit of legal wrangling (about whether the information was taken legally or not), all this information about *Paloma V* was passed on to CCAMLR and Namibia (the flag state). Namibia was ironically hosting an African conference on illegal fishing at the time. Rather than prosecuting them, Namibia simply de-flagged the vessel and CCAMLR blacklisted them. The latest word is that the *Paloma V* is still operating, though now under a Mongolian flag (that well known naval nation). This is what illegal fishers do: change their name, change the flag, set up a shell in a new country with corruptible officials and carry on fishing.

Why didn't New Zealand prosecute them through our courts? Remember that under UNCLOS, given the crime happened in international waters, the country to which the vessel is flagged has jurisdiction. Sorting this problem out takes a lot of international cooperation and pressure on countries that flag illegal vessels as well as the ports they land in. It's a lot of work, and at times, it is slow going. But things seem to be plodding slowly in the right direction. Life is slowly getting tougher for illegal fishers. There are thought to be only a handful left in action.

The international community has decided to allow fishing around Antarctica. To avoid the race for resources they have chosen to manage fishing, instead of banning it altogether as they did with mining. The science and intentions in CCAMLR are good, but the imperfect nature of international negotiations means some aspects of the management and execution have a way to go. While toothfish is still an 'Olympic fishery' there will always be a race for resources. What's more, illegal fishing is always difficult to manage, given a vast ocean and international negotiations. CCAMLR is not above the usual sort of camel trading ('scuse the pun) that goes on elsewhere. At the moment, the fishing is good as toothfish is fished down to the level of 50%. Will CCAMLR be able to rein fishing in when the target of 50% is approached? Will fishing nations agree to comply? We'll have to wait and see.

Even a managed fishery can damage the environment, so New Zealand and CCAMLR recognised the need for marine protection as well. New Zealand has been working up a proposal for the Ross Sea region which balances fishing and protection, but of course this isn't enough to satisfy some environmental groups. They have seized on issues such as the uncertainties in the science used for management to call for a marine protected area that would effectively eliminate the toothfish fishery in the Ross Sea region.

Marine Protection

In 2009, CCAMLR pledged to create a network of Marine Protected Areas (MPAs) by 2012. Early on, scientists flagged that the Ross Sea would be a priority area. The New Zealand Government, seeing ourselves as the responsible agent for the Ross Sea, undertook to put together a proposal. The United States, in their infinite wisdom, came up with a different proposal. A group of environmental groups – the Antarctic Ocean Alliance – then jumped on the bandwagon, arguing for even greater levels of protection. How much protection is enough? Let's start with why we need MPAs at all.

There is a good argument for marine reserves. As we discussed in *Hook, Line and Blinkers,* the removal of one species sends echoes right throughout an ecosystem. A useful rule of thumb is that fishing one species down to the level usually considered sustainable by most fisheries managers (20-35% of the original biomass) tends to reduce the biodiversity in an area by a third. So removing the toothfish from the Ross Sea could have much bigger impacts than just taking the fish: for example, less toothfish in the Ross Sea might mean fewer killer whales, too. Marine reserves may not be needed to protect the fish, but they do help protect the integrity of the wider ecosystem, which is part of CCAMLR's aims.

Protection is becoming more important, given the other stresses the ocean faces. As we will see in the climate change section, our oceans are warming and rising, there are changes happening to sea ice, and sea water is growing more acidic due to all the carbon dioxide it is absorbing. Shipping and tourism can bring new foreign organisms to the environment which can displace existing ones. As we will see later, greater biodiversity improves the resilience of an area to these sorts of stresses, which is where marine reserves may help. And marine reserves do seem to be more effective in areas that are relatively untouched, like the Ross Sea.

All of this is recognised by CCAMLR's marine protection objectives, which are:
- Protection of representative examples of marine ecosystems, biodiversity and habitats
- Protection of key ecosystem processes, habitats and species
- Establishment of scientific reference areas

So we clearly need marine reserves. The questions are how much marine protection is enough, and — this is so often overlooked by those dreaming the green dream — what is politically possible? The fact is that this is not New Zealand's decision to make. The legal toothfish fishery in the Ross Sea is managed by CCAMLR, so any change to the current set-up needs the agreement of all partners, including other toothfish fishing nations such as Korea, Russia, Japan, Ukraine and even the UK. It is perfectly reasonable to have a moral belief that the whole Ross Sea region should be locked up in a marine reserve, but we shouldn't expect these other cultures to share our morals. As such, the presumption has to be that fishing will happen, and debate over protection therefore has to proceed on the basis of science.

The New Zealand draft proposal as it stands is based on science, and is the result of an open and transparent process, known as 'systematic conservation planning'. First, they worked with scientists to understand what we know about the environment of the Ross Sea – such as where the important habitats are for penguin, seal and whale feeding. Then they gathered different stakeholders together, including scientists, fishers and environmental groups, and identified the habitat they thought was most important to protect. Then using computer modelling, they worked out how they could adequately protect the key elements of habitat with the minimum negative impact on the toothfish fishery. The scenario they came up with is depicted in Image #8.

The trade-offs of the draft New Zealand proposal are clear: fishing will be eliminated from 15-20% of its former range while the ecosystem features that scientists consider to be particularly important and vulnerable to the potential effects of fishing will be 90-100% protected. These are the bits that the Antarctic Ocean Alliance rightly point out need to be protected as a 'natural laboratory'. In fact, under the draft New Zealand proposal the Ross Sea proper is almost fully protected (95%). Most of the argument is actually over the area north from the Ross Sea to 60 degrees south. This is known as the 'Ross Sea region', an arbitrary political boundary rather than one that has any significance for the ecosystem.

The United States and Antarctic Ocean Alliance proposals were not developed through the same systematic process, so the trade-offs between protection and other activities, including fishing, are murkier. Under the New Zealand proposal, the protection priorities are clearly stated: it is simple to shift the proposed boundaries of the Marine Protected Area and immediately see the impacts on both fishing and the protection of important habitats. By contrast, it is difficult to tell from the Antarctic Ocean Alliance proposal which areas they consider to be most important to protect. Would they be willing to give up parts of the Ross Sea itself for some of their additional proposed areas outside the Ross Sea? Not bloody likely.

The Antarctic Ocean Alliance has claimed that their alternative is not anti-fishing, but this is sophistry. In reality, the Antarctic Ocean Alliance proposal would eliminate so much of the existing toothfish fishing grounds as to make fishing of the paltry remainder uneconomic. In fact, one of the main justifications for protecting the additional 'Area Two' put forward by the Alliance is that it is a toothfish feeding ground. It is impossible to protect 100% of toothfish feeding grounds and still have a viable fishery; where else do they propose to go fishing – where toothfish go for Ramadan?

Many of the other Antarctic Ocean Alliance arguments for their additional protected areas are questionable. They claim to be concerned about benthic habitats, yet bottom trawling is already banned in the Ross Sea region, all fishing is prohibited from grounds shallower than 550 metres' water depth (the main areas of benthic diversity are shallow) and the bottom impacts of long-lining are minimal. Some of the features they want to protect (like underwater ridges and troughs) couldn't possibly be harmed by fishing, unless fishers started using depth charges. Finally, their claims that some of their additional areas are important feeding areas for Weddell seals and killer whales are shaky, with over 90% of these areas already included within the New Zealand draft proposal.

The Antarctic Ocean Alliance campaign on marine protection is the traditional NGO stance: take the extreme position in the hope that achieves something moderate. It looks like Negotiating 101; it certainly confuses the public and risks a far more serious consequence. What is even stranger is that some of the local NGOs were involved in the New Zealand Government's process to develop our MPA scenario before they had to get in behind the Antarctic Ocean Alliance proposal, which seems to have come out of the international NGO offices. The politics the Alliance plays could embarrass Western Governments into taking a far tougher, less realistic stance on marine protection. Such a position may win politicians the Green vote, but it would be a Pyrrhic victory. A holier-than-thou stance of this nature could be immediately vetoed by fishing nations and that would derail the whole marine protection campaign in the Ross Sea region. That would be an extremely bad outcome — and all solely due to very risky tactics that sacrifice *realpolitik* at the altar of ideological piety.

Of course, the environmental movement can point to instances where their tactics have worked – such as their opposition to mining on Antarctica. In the midst of negotiations for a treaty on mining (in Wellington, no less), the idea of declaring Antarctica a world reserve caught on and ended up as a reality. So maybe idealism sometimes pays off. It's more likely that most countries agreed to this because there was no realistic prospect of mining in Antarctica for the foreseeable future. Fishing is already an established industry, and one that fishing countries won't give up so easily.

There are other, recent examples where trying to impose Western, self-righteous, selective morals on other cultures has backfired. Take the example of the International Whaling Commission discussed above. The piety preached by countries and NGOs opposed to any whaling whatsoever almost certainly pushed the Japanese into hardening their own line, and meanwhile, New Zealand's proposal — quite possibly the best chance the Japanese have ever been offered to make a graceful exit from whaling — went by the wayside. Could the same happen in the toothfish fishery?

The final pragmatic issue with the Antarctic Ocean Alliance proposal is that without any legal fishing in the Ross Sea region, the Ross Sea itself could be more vulnerable to illegal fishing. Where toothfish stocks around Antarctica have been overfished in the past, it has primarily been by illegal fishers. Having legal fishers operating just outside the Ross Sea is one of the best means of protection we have from illegal fishers getting into the Ross Sea. The legal fishers are always pretty quick to detect and report the bad boys and that kicks the international machinery of sanction into action. That's one reason having no legals down there could mean open slather. The other, of course, is that if we're not getting anything out of the toothfish fishery, what incentive do we have to contribute to the protection of the stock — how does the government justify all the expenditure on Orion overflights, patrols by offshore fisheries patrol vessels and so on? It's perfectly analogous to the philosophy of the Department of Conservation toward the conservation estate. If people can't use it, they won't value it, and they certainly won't want to pay for it to be preserved.

There is nothing wrong with believing that everything south of 60 degrees South is sacrosanct and should be locked away in a marine reserve, except that nobody has the unilateral power to enforce that high ground. These are international waters, and the Convention governing them allows for 'rational use' of the fish resource. This is a great opportunity to make real gains in protecting the environment and its species. We must be mindful of the realities of international negotiations faced by our government. If they manage to achieve anything like the current proposal in the Ross Sea, it would be a huge success. We should pat them on the back and tell them to get on with pursuing a similar process in our own waters. The Antarctic Ocean Alliance strategy, on the other hand, risks a repeat of the balls-up that has occurred with whaling.

This debate over marine protection in the Ross Sea threatens to overshadow another, arguably more urgent problem here – the lack of marine protection in New Zealand's own waters. It would be far easier to create and police substantial marine reserves in our own EEZ, rather than in the Ross Dependency. New Zealand is a long way from having a comprehensive network of Marine Protected Areas in our own EEZ. We don't even have a set of objectives – so that should be the priority, rather than relying on imperfect international processes to put a figleaf over our shortcomings.

While New Zealand has 7% of our territorial seas (out to 12 miles offshore) in Marine Reserves, the vast majority of this is in the reserves around the Kermedecs and Auckland Islands. Around mainland New Zealand, only about 0.2% of our territorial sea is protected, and the Government has called a halt to the creation of any more. Some marine reserves around the subantarctic islands are planned: these were the result of a fraught process that pales in comparison to our efforts in the Ross Sea region. We don't even have the legal ability to declare marine reserves in our EEZ beyond the 12 mile limit, thanks to our outdated marine reserve legislation. Clearly we are not as clean and green as we like to tell the outside world we are.

Summary

The exploration of *Our Far South* was part of a race for resources and territory that continues to this day. Thanks to our rapacious sealing, whaling and farming in the subantarctic islands (a legacy from which they are still recovering), New Zealand was able to secure sovereignty over those rocky isles. This in turn gained us one of the largest areas of EEZ in the world.

The race for territory in Antarctica was more hotly contested, and in the end the nations with an interest in the area decided to sign a Treaty which buried the issue by agreeing to disagree over sovereignty. This Treaty has proved incredibly successful at ensuring the continent is dedicated to peace and science. This is in our interest: we are just too small to get into a turf war.

However, over time, cracks may appear in the Treaty system. The issue of jurisdiction was never resolved, which means that any issues that arise between nations need to be resolved diplomatically. And any disputes could threaten the integrity of the Treaty system. The Treaty may have frozen the status quo from the 1950s, but in reality, the balance of world power is changing. We are seeing this as emerging nations flex their muscle by building stations and increasing their presence in Antarctica, under the customary guise of science. And the potential for a race for resources has been ever-present, with a constant tension between totally locking up the resources so no treaty member can use them, or managing the worst effects of commercial exploitation.

Antarctica appears to be an untouched wilderness, but there is already commercial exploitation and environmental damage, particularly of fish stocks and on the relatively rare ice-free coastal sites where scientists and tourists like to hang out. While mining and drilling may one day be added to this list, it's unlikely in the foreseeable future because of the Madrid Protocol and (probably more importantly) because the economics don't stack up.

The realpolitik is that the race for resources will never completely disappear. In the vagaries of international negotiations, there is no way to call a full stop to countries exploiting the Continent. In fact, what has already been achieved in Antarctica through the Antarctic Treaty System easily outstrips what has been achieved in most other international talkfests: just look at climate change. The question is: how can we as a country work to keep the Antarctic Treaty system effective, credible and relevant? Certainly not by unilaterally declaring rules and regulations in some silly, vain belief that anyone else will take notice. We have to be a whole lot smarter than that. We have to manage the race for resources. This will involve striking a balance between complete protection of the area and managing the worst aspects of commercial exploitation by agreeing environmental standards and setting aside some areas as complete reserves.

This is ultimately a diplomatic exercise. We are constantly walking a tightrope between involving more countries to stop them doing their own thing, and how much it is worth compromising our values to involve another country in the Treaty system. The recent admission of Malaysia (discussed below) is a good example. But this diplomacy is worth investing in because overall New Zealand has hugely benefitted from having an Antarctica dedicated to peace and science. It helps keep our little corner of the world nice and quiet.

The Admission of Malaysia

For many years, Malaysia has been the voice of dissent in Antarctic circles. They have labelled the Treaty system a rich white man's private club, and have led the call for a system that involves more developing nations. They argued that Antarctica is the 'common heritage of mankind' and should be run by the UN.

While it is true that the Antarctic Treaty System doesn't include all nations, there are reasons. Conservation of Antarctica is a key value for many members, and many countries outside the Treaty system would probably want to focus more on resource use.

New Zealand has long worked with Malaysia to show the value of preserving the continent for peace and science. Our aim has been to get them inside the tent working with us rather than leaving them outside, being disruptive. This culminated in their admission in 2011. Malaysia now has a fairly modest scientific programme in place (which operates out of Scott Base and among other things involves trying to grow tropical plants in the cold) and is on the way to becoming a Consultative Party.

The question we really have the power to ask ourselves as Kiwis is whether we are doing the best we can within the area that we actually control: our own EEZ. We could be doing a whole lot more to protect the ocean and unique life on the subantarctic islands. But this isn't a priority for our resource-hungry government. Some funding-driven NGOs also seem to find it sexier to grandstand over the so-called 'pristine continent' than to focus on our lack of domestic marine reserves.

CLIMATE CHANGE

Perhaps the most astounding feature of climate change is that there still seems to be debate in New Zealand about whether it is real and whether humans have caused it. The fact is, we already know a lot about what's happening, and what to expect. The debate should be about what we're doing about it.

Nothing is ever certain in science; theories can be changed and overturned. However, as we saw in *Poles Apart*, written with John McCrystal, the vast majority of evidence supports the theory that human emissions of greenhouse gases are impacting on the world's climate, and the theory itself is pretty well unimpeachable. And yet the most recent surveys suggest that around 40% of New Zealanders either doubted or denied that human actions are adversely altering the climate.[35] By contrast, amongst the international community of climate experts, the percentage of doubters is only 3%.[36] Too often the land of talkback radio and public opinion reminds us of religion – 92% of Americans believe in God, 40% of scientists do and only 7% of eminent scientists do.[37]

This is an indictment of the media, through which most people get their information. Whenever discussing climate change, the media are careful to give voice to climate change deniers. As a result, a tiny minority of the scientific community manage to get half the column inches. It really is an odd state of affairs, and one that needs urgent correction. Of course, there will always be those who disagree with the research because it doesn't suit their purposes. Sadly such red-necks will never be persuaded, at least until the waves are lapping at the doors of their bach. By then it will be too late.

For the benefit of those still being suckered by soundbites from provincial media, or who haven't had time to get informed on this subject, we'll begin with a quick survey of the natural phenomena that scientists know have an influence on climate and consider their likely contribution to the climate change we're experiencing. Readers interested in getting into the detail need to read a survey of the science such as the one we presented in *Poles Apart*.

Then we'll look at the evidence of the current climate change's impact on *Our Far South*.

Climate Change 101

Scientists — and even most deniers — agree that warming is happening; the only questions are what is causing it, and to what extent is it human influence? There are several forces that affect our climate over different time scales and the key to understanding climate change is knowing what these all are and what contribution each is making. It's only after performing that logical calculation that it becomes clear that what we're seeing is not explained by the natural phenomena that have driven and still drive our climate changes.

Over millions of years the movement of tectonic plates changes the ocean currents which in turn affect the distribution of heat around the planet. Over time scales of tens of thousands of years, changes in the geometry of the earth's spin and orbit around the sun affect incoming radiation. Over much shorter time scales – years to centuries – climate is influenced by volcanic eruptions and variations in the sun's output, known as 'sunspots' (although both these factors are short lived). An appreciation of the natural phenomena that have influenced the past – distant and near-term – is essential to recognising that no combination of these factors can explain what we're experiencing now.

In the early part of the twentieth century, a Serbian engineer named Milutin Milankovic tried to solve the mystery of Earth's past ice ages. He noticed regular patterns in the freeze-thaw cycle through history, and matched these with the observation that the Earth's orbit around the sun is not constant and that there are eccentricities in the way Earth spins. As the Earth rotates around the sun, there are regular and predictable changes to the way the Earth wobbles (every 21,000 years) tilts (every 40,000 years) and even the path of the orbit itself (every 100,000 years). All of these factors influence how much sun hits the planet, and where. Thanks to Milankovic's calculations, we know that these changes are very predictable throughout time, and we can therefore take them into account. These changes are very slow compared to human lifetimes and civilisation and do not explain what we're experiencing now.

Sunspots are a sign of increased activity on the sun, which in turn has an effect on the amount of radiation that hits our atmosphere; these periods of heightened activity may cause minor warming. In other words, you can think of sunspots as being like someone just threw another log on the fire. Again, fluctuations in activity seem to happen in fairly regular cycles (around every 11 years) and they are easy enough to monitor. Importantly, they are cycles – the effect comes and goes: it does not explain the sustained trend in temperature we're experiencing. Although the cycle is quite regular, the height of the peaks is quite hard to predict, and for unknown reasons, they stopped altogether for a while during the 17th century. Quite separate from sunspot activity there are actual changes in the amount of energy the earth receives from the sun. These are quite small, but they go on for a long time, and so have a measurable effect on our temperature. Over the last 50 years, the trend has been for the sun to get slightly weaker, which ought to be having a cooling effect. But as we know, earth's temperature has risen over that time.

Volcanic eruptions can also have an impact on the planet's climate. Really big explosions send ash and sulphur dioxide into the stratosphere (really high in the atmosphere), where it stays for a while and spreads right around the world. These particles scatter incoming solar radiation and deflect much of it from getting in. Of course, volcanoes are unpredictable, but when they go off we can see their impact in action. In 1991, Mount Pinatubo in the Philippines blew its top in a big way, pumping an estimated 20 million tonnes of sulphur dioxide into the atmosphere. Some of this got high enough to reduce global temperatures by between 0.3-0.5°C.[38] Volcanoes have a big cooling effect, but that effect only lasts a couple of years, so the effect on earth's average temperature really depends on how often we get a big eruption.

Finally, we come to greenhouse gases (carbon dioxide, water vapour, methane, and nitrous oxide are the main ones). Of course, almost all energy in our climate comes from the sun. Greenhouse gases help our planet hold on to the solar energy, keeping the planet warm. They work a bit like this. The sun radiates energy as light and because our atmosphere is pretty transparent to light, about half of it gets through our atmosphere and is absorbed by the land and ocean. The earth has to find a way to release this energy otherwise it would keep heating up and temperatures would soar. So the earth radiates energy back into the atmosphere and out into space as infra-red light; we can't see infra-red radiation, but we feel it as heat.

Carbon dioxide and other greenhouse gases act like a smoggy haze for infra-red radiation (this is thermal radiation and is not to be confused with the dangerous high-energy nuclear radiation). In short, energy can get in through the atmosphere, but much of it is trapped inside the atmosphere on its way out. One way for heat to escape is through rain: water absorbs heat at ground level in the process of evaporation and then releases it when it condenses high up in the atmosphere, forming rain (or hail, or snow). The heat thus shifted higher up in the atmosphere has a greater chance of escaping into space. Meanwhile, the heat has been trapped in the atmosphere, warming the earth up. This isn't all bad news: without any greenhouse gases, the average annual temperature on Earth would be a nippy – 18°C instead of the rather mild +14°C it is now.[39]

We know the world is warming, but which of these factors is the primary cause? Luckily we have plenty of historical data to look at so we can work out how much impact each factor has. We have data on the earth's temperature and atmosphere over millions of years from cores taken in ice and deep ocean mud. The Milankovic cycles are regular and very slow, so scientists can account for their effect on temperature. We have data on solar activity going back hundreds of years and there is some regularity to this as well, making predictions of the timing and impact of these natural cycles straightforward. Volcanic activity can also be accounted for, using historical observations and ash deposits.

Once you strip out all these effects, the impact of greenhouse gases on the climate is very clear. The only models that can explain the recent rise in temperatures are the ones that include an allowance for man-made greenhouse gas emissions. The recent rise in greenhouse gases is having exactly the impact on our planet that you would expect from experimental physics.

Like the other greenhouse gases, carbon dioxide has a strong effect on our temperature even though it's so rare in the atmosphere that it is measured in parts per million (ppm). In the million years prior to the industrial revolution, levels of CO_2 in the atmosphere fluctuated between 180-280ppm (according to measurements made on air bubbles trapped in ice cores). Concentrations of atmospheric CO_2 have been directly measured since the 1970s; New Zealand's Baring Head (opposite the entrance to Wellington Harbour) has the second longest direct record of atmospheric CO_2 and methane, and the oldest in the Southern Hemisphere. In 1970, when these measurements were first taken, levels stood at 324ppm; the most recent reading was almost 390ppm, almost a 40% increase on maximum pre-industrial levels.[40]

What is most worrying is the speed of the change of temperature. In the time that homo sapiens have walked the Earth, there have been fairly large temperature swings, across a range of 2°C warmer to 8°C colder than the present; but these have happened slowly enough for nature to adapt. The pace of the current temperature change is a different proposition. The Earth shrugged off the last ice age by warming up at an average of around 0.005 degrees per decade. At that rate a 2 degree rise in global temperatures would take 4,000 years; yet this is the temperature rise that the Earth is likely to experience just in this century alone (if not more). According to the IPCC (the Intergovernmental Panel on Climate Change – kind of a Jedi Council of climate change scientists) *"there is no evidence that this rate of possible future global change was matched by any comparable global temperature increase of the last 50 million years"*.[41]

Partly thanks to climate change and human development more broadly, rates of species extinctions could be 100 times (some estimate up to 1,000 times) that of the natural rate according to the fossil record and this is projected to increase by a further factor of ten in the near future.[42] If nothing changes, then the fossil record of the Anthropological Age may resemble another asteroid hitting Earth, such as the one that brought down the curtain on the dinosaurs. Given the risks we face the longer we delay, this warrants some action to cap the stock of long-lived greenhouse gases in the atmosphere. Reducing our emissions is the most obvious way to do this. So far, there is no sign of any ease-off in the relentless rise of CO_2 concentrations or of temperatures, as shown by the graph below.

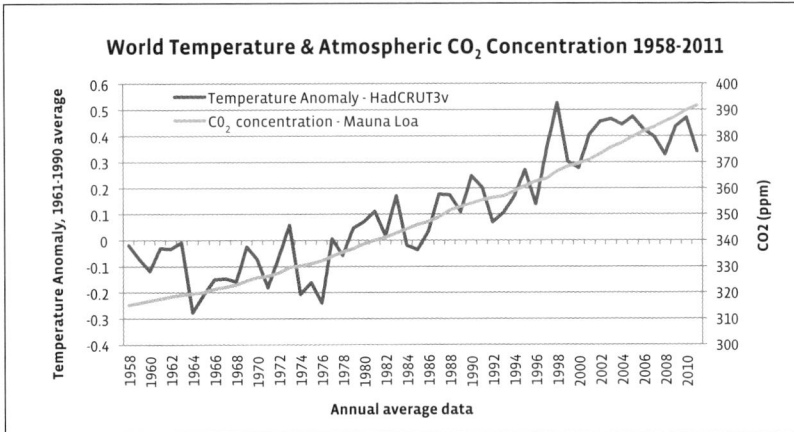

World Temperature & Atmospheric CO_2 Concentration 1958-2011

Temperature Anomaly - HadCRUT3v
CO_2 concentration - Mauna Loa

Annual average data

Sources – National Oceanic and Atmospheric Administration, US Dept Commerce and Met Office Hadley Centre for Climate Change

There are other signs that things are different this time. In the past, warming has preceded rises in carbon dioxide levels, believed to be the consequence of warming the oceans and reducing their capacity to absorb their share of natural emissions. This fact has been used by deniers to argue that there is no causal connection between CO_2 rises and temperature, and that their paired rise in modern times is merely a correlation. But this is a simple fallacy. Just because a temperature rise historically precedes an increase in CO_2 levels doesn't mean it necessarily does. That is the natural order, after all, and the present warming is not natural. Nobody is disputing that we are pumping previously locked up (or "sequestered") carbon dioxide into the atmosphere. Besides, the effect of the rise of CO_2 levels subsequent to an initial warming has been to amplify the warming,[43] and we know from isotopic analysis of the extra carbon up there that it is coming from fossil fuel combustion.

Never before has the question had to be asked: what are the consequences of pumping CO_2 relentlessly into the atmosphere?

What will happen in the future? We have a general idea (and it doesn't look good), but a lot depends on how much we emit and exactly how the planet reacts. While the planet has some self-regulating mechanisms that might take the edge off climate change and keep things bearable, there are also feedback loops which, once triggered, could make things progressively worse. A lot depends on how these two forces balance each other out in the future. Scientists are busily studying the past to unlock the secrets of what the consequences will be of this human-induced warming and at the same time they are closely monitoring current trends. What they're no longer doing is asking the question whether the observed warming is human-induced or not. That remains the preserve of the least-informed.

Our Far South has a unique role to play in the story of climate change. First, the polar regions are more sensitive to some types of change, so they act as a barometer for the changes to come (see Image #9). Second, *Our Far South* is one of the most untouched places on the planet. In particular, the ice covering Antarctica, accumulated over hundreds of thousands of years, provides us with a unique opportunity to study both the current and past climate from the environmental signals trapped in cores of ice and mud. It's been called a 'climate museum' for this reason. Finally, as we saw in the first section, *Our Far South* is one of the main drivers of the global ocean and climate. The influence of the ACC and the sea ice freeze/thaw cycle are felt around the world. Changes in any of these things could create big issues for the planet.

The best place to start is to look at what is happening now and what that might mean for the future. But remember, it's early days in climate change; reliable records of Southern Ocean waters go back only 60 years. So as a guide to what might happen in a warmer world, it also pays to look at the distant past to get an idea of how the ocean and continent have behaved. Finally, we'll look at what all those possible futures mean for us here in li'l old New Zealand.

What Can We Learn From Current Trends?

Over the past 100 years, the atmosphere has warmed on average by around 0.7°C.[44] However, it's the global ocean that is actually bearing the brunt of climate change. The upper layers of the ocean have warmed by 0.6°C since the 1950s,[45] but because of its large thermal mass (particularly in the deep ocean) the ocean stores a lot more heat during the process and it is slow to let it go. In all, it appears that the ocean has in the past been storing up to 90% of the increased heat from human-related climate change.[46]

Since the amount of fossil fuel consumed from year to year is known with a reasonable degree of precision and direct measurements of atmospheric CO_2 yield a figure for the amount of carbon reaching the atmosphere, the difference – the missing fraction – must be retained by the oceans. As we have seen, the ocean stores some 30% of the carbon dioxide emissions we make. But this ability to absorb carbon dioxide comes at a price, namely increased acidity in the ocean, which we'll look at shortly. And as the ocean warms, it is losing this ability to absorb gases, in much the same way Coke fizzes more energetically as it warms when you take it out of the fridge. The molecules in warm water have more energy, so they move more and there is less room between them to hold gas molecules, so the gas gets bumped out. In 2009, the world's oceans only absorbed 26% of our carbon dioxide emissions; is this a sign of things to come?[47]

But for those of us with a bach on the coast, perhaps the most worrying sign is the slow but steadily increasing rise in the sea level. Since 1950, the sea level has risen at an average rate of about 1.7mm per year, but since 1993 the average rate has increased to over 3.1 mm per year. [48] This might be too small to notice in one year, but the impacts are starting to take effect on our coastline, particularly during storms.

This increase is due to two factors. First, *thermal expansion* – the fact that water expands and takes up more space as it gets warmer. And second, the additional water from melting glaciers and ice caps, particularly in temperate regions. Thermal expansion accounted for the lion's share of sea level rise in the 1990s, but since then, increased melt from high altitude glaciers has taken over. The really big polar ice sheets in Greenland and Antarctica haven't contributed much to the melt so far, but their contribution is increasing.

The warming trend has been more noticeable in the Arctic than anywhere else, where it has been getting warmer at twice the rate of the rest of the planet. This is partly because it is surrounded by lots of land (which heats up more quickly), but also because of the albedo effect we talked about in the first section — the loss of sea ice means that less sunlight is reflected into space, so the land and seawater exposed absorb more energy, which is converted to heat. Not surprisingly, then, the Arctic is rapidly losing sea ice and if the current rate of decline is maintained, in around 30 years the North Pole will be surrounded by a mixture of open ocean and patches of thin ice in the summer. Remember melting sea ice doesn't change sea level.

While the evidence of climate change across the world is compelling, sorting through the visible symptoms in *Our Far South* requires more than a cursory glance. There is incredible regional variability, but overall there is a warming trend. We'll start with the Southern Ocean, before moving on to Antarctica.

Southern Ocean

Some of the strongest evidence for climate change is in the ocean. The Southern Ocean has become warmer, fresher (less salty) and slightly more acidic and the patterns of circulation have changed. As we will see later, these changes have been linked with population declines of some animals in our subantarctic islands.

As we saw in the first section, the Antarctic Circumpolar Current (ACC) is the engine room of the Earth's oceans and climate. The incredible churn of water that occurs increases the amount of water that comes into contact with the atmosphere, increasing the amount of heat and carbon dioxide transported to the ocean's depths. This has mitigated the impact of global warming, particularly in the Southern Hemisphere.

Throughout most of the ocean, only the surface layer absorbs heat and carbon dioxide to any great extent, but in the Southern Ocean, both are being stored down to depths of 1,100m. As a result, some 40% of carbon stored in the entire global ocean is south of 30°S, mostly between 30-50°S,[49] and the waters of the Southern Ocean are warming more quickly than the global average. The upper 700m of the Southern Ocean has warmed by 0.2°C since the 1960s.

How do we know all this? European-based and focused scientists used to think it was the annual freeze and thaw of sea ice around Greenland that helped drive the major currents that arose in the North Atlantic and passed through the world's oceans. Thanks to a nifty invention called the Argo float (more on these in the box on the next page) we now have much better information on the global ocean. As usual, more information makes things more complex; we have learned that global currents do not behave as a uniform flow. What we do know is that there are many drivers of currents, including sea ice, ocean density and eddies; but the main force behind them is global wind, which has brought the ACC into centre stage.[50]

Argo Floats

Over the past few decades we have learned more about the importance of rising sea levels and the role of ocean currents in our climate. At the same time, we have increasingly realised how limited that knowledge of the ocean has been, particularly the deep ocean. Ocean science was difficult – the closest analogy would be trying to study the Earth from a spaceship parked above a thick layer of cloud. The results gleaned from the Argo float programme in the last decade have changed all that.

Prior to Argo, studying changes in the ocean was difficult and expensive. Satellites could collect continuous worldwide data, but only of the ocean surface. For the deeper ocean, scientists relied mainly on data from research ship voyages which only provided snapshots of transects of the ocean, and only certain parts of the ocean at that. This was complemented by measurements from long-term ocean observatories, but these are few and far between. Hence, the deep ocean was not well understood.

The Argo float programme was designed to correct this lack of data. Working together, countries from around the world share information from a system of almost 3,500 'floats' that are deployed around the world's oceans. Each float sends back real-time data via satellite on ocean temperatures, salinity, oxygen, winds and currents from the upper ocean to water depths of typically 1,000m. Ships can simply drop an Argo float and move on, rather than having to stay in the one place painstakingly collecting measurements. Nor does it require a specialised research vessel to do perform the drop: a cargo vessel, a warship, a fishing boat, or a tourist vessel would do just as well. This allows for far greater coverage of the world's oceans.

The floats themselves are an impressive piece of kit, each valued at US$30,000. They look a bit like a big SCUBA tank. They have two oil bladders, one internal and the other external, and they can pump oil between them in order to alter its density and their relative buoyancy which allows them to hover in the water column at specific depths. Generally speaking, they take about 10 hours to sink to a depth of about 1000 metres, where they stay and drift for 8-10 days. Then they sink to 2000 before slowly ascending again, profiling the water for temperature and saltiness all the while. Once at the surface, they beam all the information they have been collecting to a satellite which relays it to one of the two Global Data Assembly Centres in Brest, France and Monterey, California.

On our voyage south, we released 21 such Argo floats to bolster the array in the Southern Ocean (which you can see in the map on the next page). These next-generation floats have new technology that allows them to cope with sea ice (they sense when there is ice overhead, and delay their return to the surface until it is clear) and the shallower waters of our continental shelf. Unfortunately, they don't yet work in the domestic bathtub.

Argo Floats in the World Ocean

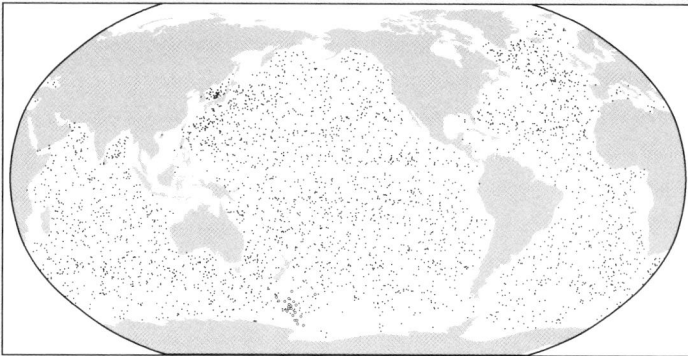

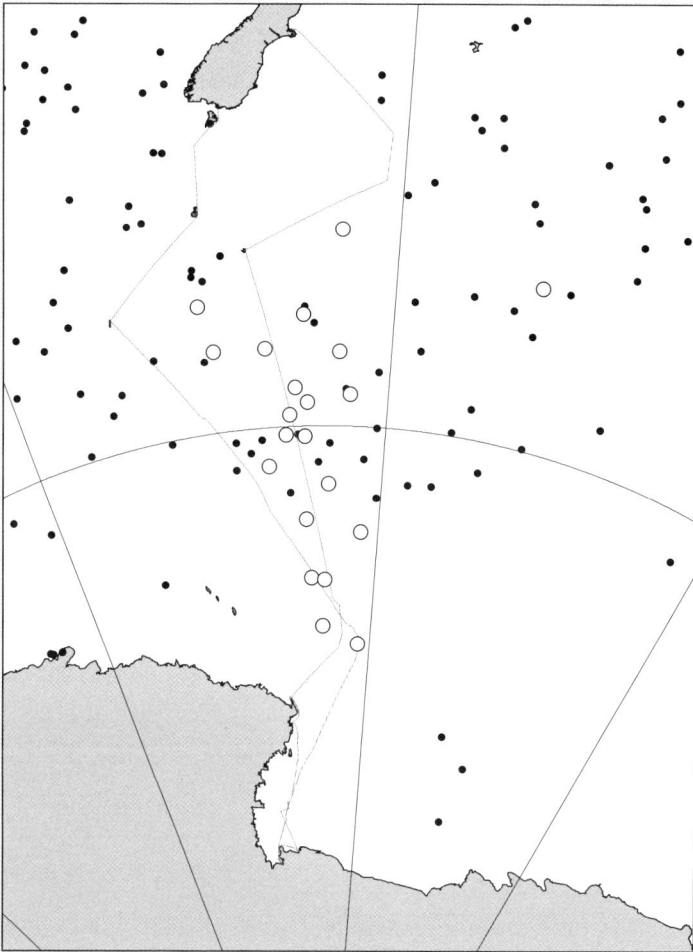

Credit: Dan Zwartz. Source: http://www.argo.ucsd.edu/. Round circles are the floats dropped by the OFS voyage (with the grey line charting the voyage itself).

Climate change also seems to be changing the workings of the ACC itself which, given its importance, requires close monitoring. So far, there seem to be two major impacts: the location of currents is changing, and the current is becoming either faster or more turbulent (or both).

As we saw in the first section, surface currents are driven by winds. Winds in turn are driven by air moving between areas of different temperature throughout the world. In the Southern Hemisphere, climate change (and the hole in the ozone layer – more on this later) seems to be increasing the temperature difference between the equator and the Pole all year round, making it windier. All-in-all, the wind speeds over the Southern Ocean have risen by 15-20% since the late 70s.

Because there is a permanent temperature difference between the temperate zones of the Southern Hemisphere and the Pole, there are always strong winds blowing somewhere in the Southern Hemisphere: the only question is where. The Northern Hemisphere is different because the Arctic warms up more during summer, reducing the temperature gradient and endowing the north with better defined seasons. Not only are these Southern Hemisphere winds getting stronger, but they are migrating further to the south, too. The area of strong westerly winds normally lies either over southern New Zealand or the Southern Ocean. As the climate is warming, the area of strong westerly winds seems to be spending more time sitting over the Southern Ocean.

All-in-all, this spells mixed news for New Zealand. The outlook is for less wind generally, but when it is windy, those winds will be stronger. Wellington readers will be asking whether that is really possible?

But what does it mean for the ACC? Satellite observations have revealed a slight shift in the ACC to the south, but movement further southward may be restricted by underwater topography. Other currents are certainly changing. Between 1944 and 2005, records suggest that the regional waters off Tasmania have warmed by a whopping 2°C. This is the result of the East Australian Current — the same EAC that the turtles ride in *Finding Nemo* — moving 350km south, thanks to the same change in wind patterns that is having an impact on the ACC.

In sum, the Southern Ocean is in the front line of climate change. It has been warming, not only on the surface but to significant depths. And it has certainly locked away more than its share of heat and carbon dioxide. The winds over the Southern Ocean are increasing in strength and shifting south, but we are unsure of its full impact on the Antarctic Circumpolar Current. Will it get stronger and/ or more turbulent? Will the current itself move further south? And what will this mean for role the ACC plays in the planet's ocean mixing? It is difficult to tell for sure; and until we know, the ACC requires close monitoring.

As we have seen, the Southern Ocean is storing heat and carbon dioxide, both of which are moderating the atmospheric and terrestrial impacts of our carbon emissions. But this comes at a cost, and it is worth looking at what we know about these side effects.

Acidification

While proving the link between burning fossil fuels, CO_2 emissions and global warming can be incredibly tricky, the direct impact of CO_2 emissions on the ocean is much simpler to illustrate. As we saw in the first section, the ocean can happily absorb a third of the carbon dioxide we emit every year, but it becomes more acidic as a result. Thanks to the churn of the ACC and the cold water, the Southern Ocean is absorbing more carbon dioxide than anywhere else. As a result, it is also becoming slightly more acidic than the rest of world's oceans. Acid tends to eat away at stuff, and in the ocean, this tends to be something known as carbonate.

Carbonate is important in the marine ecology because calcium carbonate is used as a building block by most marine life that requires a shell. This includes coral, shellfish, even the minute shells of the class of phytoplankton called *coccolithophores* (it literally means 'stone shell bearers'). Calcium carbonate is the same stuff that constitutes school chalk, or the limestone in our own Waitomo Caves: both started out as calcium carbonate laid down by shell-building organisms in the ancient ocean, although we don't recommend writing on a blackboard with a stalagmite. For shell-builders to make and maintain their shells, there has to be enough calcium and carbonate in the water to enable them to combine. Thanks to acidification, carbonate levels are falling.

The ocean has absorbed large amounts of carbon dioxide in the past, and it has a natural way of dealing with it. Remember that there are stores of carbonate in the deep ocean in the form of shells and skeletons of creatures that have drifted down from above when they die. Over time, water that is depleted in carbonate (either naturally by creatures extracting it to make shells or as the ocean absorbs carbon dioxide) cycles through the deep ocean current and replenishes its carbonate content from these deep ocean stores. Trouble is, we're emitting carbon dioxide too fast, and the oceans are absorbing so much of it, the deep ocean currents can't release carbonate fast enough. [51] Apart from a few places, such as in the Southern Ocean and North Atlantic, it takes hundreds of years just to mix surface water with the next layer down, and thousands for it to get to the real depths.

Eventually, the levels of carbonate in the Southern Ocean will probably fall below that level necessary for calcium to combine with it, and corals, shellfish and even some plankton will find it harder and harder to make shells and skeletons. This puts the organisms under stress, reducing their ability to feed, grow and reproduce because they are devoting more energy to growing their shells. The logical end point, if ocean acidity rises sufficiently, is for existing structures to be eaten away. This means degraded coral reefs, or pretty much anything with a shell. When you realise how much of life in the ocean has a shell, this is a worrying prospect.

We are already seeing signs of this in action. The shells of modern plankton called foraminifers are becoming lighter than those of their recent ancestors – and this has happened in direct proportion to the increase in acidity.[52] These little guys, no bigger than a grain of sand are a substantial part of the base of the food chain in subantarctic to tropical waters.

Changes in acidity are measured on the pH scale. This measures the number of hydrogen ions around, which in turn are a product of the reactions we associate with acid. It is a scale from 0 to 14 where 7 is neutral, below 7 is acidic and above 7 is alkaline. The scale is logarithmic, which means that a step of one actually signals an increase or decrease in acidity of 10 times. Pure water has a pH of 7, soda water has a pH of 4, while the anti-acid concoctions you take after too much Christmas lunch have a pH of 10. Before the Industrial Revolution seawater probably had an average natural pH of around 8.2, although then and now, the exact number varies depending on location. Since the Industrial Revolution, the pH has dropped 0.1, and some scientists predict a fall of 0.4 by 2100. Remember, it's logarithmic scaling, so we already have 30% more acidity (measured in the number of hydrogen ions), and we are headed for a 150% increase.

This would be a bigger change in acidity than anything seen in the past 20 million years, as far as we can tell from pH measurements on ancient rocks.[53] Like much of the rest of the ocean's response to change, such an outcome would be irreversible in our lifetimes.[54] But what would be the impact on wildlife?

In the lab, it is clear that acidification has an impact on any organism with a shell – including cold and warm-water corals, small zooplankton called pteropods, phytoplankton with protective shells and, of course, shellfish. In all cases, their growth slows, and their larvae struggle to thrive.

However, when it comes to the real ocean, we aren't sure what the impacts will be. The species that are likely to be affected first are those that rely on a particular form of calcium carbonate, known as aragonite, which is more sensitive to change in acidity. Aragonite is used by corals and some creatures such as pteropods. These beautiful but tiny animals (their name means: wing-limbs: thanks to wing-like structures, they look like miniature angels or butterflies) are part of the nutritious soup sloshing around in the sea known as zooplankton. Together with phytoplankton, they are the first rung on the food chain, munching on phytoplankton and in turn getting munched by pretty much everything else – including the larvae of many growing fish. Thanks to its cold waters and the churn of the ACC, by the end of this century the Southern Ocean is expected to be the first area of surface ocean that will not have enough aragonite for pteropods to make shells.

The good news on acidification (if you can call it that) is that, as we have seen, the rate at which the ocean is absorbing carbon dioxide seems to be slowing due to the water warming. The downside of this is that it is the equivalent of the oceans chucking in the towel on carbon emissions, and telling us we are on our own. If less is absorbed by the oceans, CO_2 emissions will from that point have an even more direct and immediate impact on the atmosphere and our climate.

In short, the warning signs of acidification are pretty ominous. Along with climate change this is one hell of a risky experiment we humans are embarking on with our oceans. The Southern Ocean, in particular, is absorbing more than its share of CO_2 because it is colder *and* can transfer CO_2-rich waters to depth. This is great for reducing the impact of our emissions, but not so great for the ocean because it is becoming more acidic. As far as marine life goes, Antarctica is a food bowl, and making life harder for creatures like pteropods is a bit like putting weed-killer on our best pastures and wondering why the cows and sheep die. At least nowadays the impact of acidification on plankton and shellfish in the Southern Ocean is being closely monitored by our scientists (see the box below).

NZ Acidification Research

The issue of acidification is relatively new to science – the first science publications only started appearing 10 years ago. Understanding the problem quickly is critical, and Kiwi researchers from NIWA and Otago University are certainly doing their bit to understand how the issue will play out.

In a purpose-built Wellington lab, NIWA scientists have been testing the impact of the level of acidification predicted by the end of this century on Antarctic organisms. While their experiment didn't result in any mortality, the organisms studied all experienced greater stress from the higher acidity levels.

The laboratory is one thing, but reality can be far more complex. NIWA have also been collecting and monitoring the types of plankton on the Chatham Rise. They are doing this in two places – north and south of the Subtropical front. They are finding that in the cooler, subantarctic waters south of the subtropical front, quantities of pteropods are falling. Is this the canary down the coalmine of ocean acidification?

Antarctica

The Arctic, and its polar bears in particular, are the poster child of climate change. Sea ice up there has been declining by a massive 12% per decade since satellite records began in 1979. In 2011, the area normally covered by older, thicker ice was 60% smaller than it was in 1981.[55]

By contrast, signs of the impacts of climate change in Antarctica have been a little more equivocal. For starters, the measurement of temperature is patchy on Antarctica. The quest for temperature data almost cost the American explorer Richard E. Byrd his life when he nearly died of carbon monoxide poisoning from his stove while wintering alone in the middle of Antarctica in 1934. The oldest consistent temperature records, those from the Antarctic Peninsula, go back only to 1951, although records from the nearby Orcadas Islands began in 1904.

From the data we do have, surface temperatures are extremely variable across the continent. This should really be no surprise, because it is a massive expanse that is covered in ice sheets up to several kilometres thick, has a suite of microclimates and is surrounded by the Antarctic Circumpolar Current, which seems to have an insulating effect.

There is a slight positive trend in temperatures, on average around 0.2°C in total, since the late nineteenth century. This unspectacular average result conceals huge variation across the continent. Things have warmed the most on the Antarctic Peninsula, by up to a whopping 0.56°C per decade since the 1950s. That is the change in annual temperature, but if you look at winters only, the change is 1°C per decade. That is the greatest change of any part on earth, apart from the Arctic. Over West Antarctica as a whole, temperatures have risen a moderate 0.1°C per decade, while in the centre of the high Antarctic plateau, temperatures have actually dropped.

In part, it is thought that this anomaly in the centre of Antarctica has to do with another major issue affecting the atmosphere in *Our Far South*: the 'hole' in the ozone layer. In case you slept through the 1980s, or were too busy listening to Duran Duran and Aha! to notice the biggest environmental issue of that decade, there is more information on the hole in the ozone layer in the box on the next page.

Holey Ozone Layer, Batman!

Ozone is an unusual molecule. Normally oxygen has two oxygen atoms bonded together, but in ozone we find three oxygen atoms bonded together. This molecule hangs out in high concentrations high up in our atmosphere and helps to protect us from ultraviolet radiation from the sun (the stuff that can burn your skin) by turning it into heat. Kiwis know about the depletion of the ozone layer only too well. Many of us have had to adopt the slip-slop-slap mentality during our lifetime to avoid turning into a 'lobster' (as the early adverts called it) which might lead to harmful skin damage and even cancer.

The enemy was chlorofluorocarbons (CFCs), which were used in cooling units and as propellants in aerosol sprays until we got wise to the problem in the 1980s. It is the chlorine in the CFC that really causes the problems, but normally chlorine gets destroyed before it reaches the stratosphere. The CFC molecule protects the chlorine until it gets into the stratosphere where it breaks off and acts as a catalyst, turning ozone back into oxygen at an alarming rate. One chlorine atom alone can destroy up to 100,000 ozone atoms.

CFCs have caused a slight depletion in the ozone layer all over the planet – between 3-6% in most places. But over Antarctica the process has gone crazy – resulting in a complete ozone hole. Why has Antarctica been so different?

It turns out the extreme cold of Antarctica is the cause of its downfall. Once temperatures hit 80 or 90°C below zero, icy clouds form in the stratosphere. These clouds freeze some other chemicals (including nitrogen dioxide) which usually help to keep chlorine in check. In their absence, chlorine has free rein to destroy as much ozone as it can. This is why the ozone hole appears over Antarctica in winter of every year, and slowly repairs itself in spring and summer.

Image #10 shows the largest ever ozone hole over Antarctica in September 2006. The purple and blue colours are where there is least ozone, and the yellows and reds are where there is more ozone. The hole matches almost exactly the area of the polar vortex – the area of ultra-cold air over Antarctica.

Unlike on climate change, governments and industry acted quickly and decisively on CFCs, probably because they had an easy alternative to switch to. In 1989, the Montreal Protocol banned the use of CFCs. The total amount of ozone-depleting gas has dropped by about 10% since then, and is now falling by about 1% per year. Scientists think that the Antarctic ozone hole has probably stabilised, and from now they expect the ozone to slowly regenerate over time. But it's still there every spring, so don't go putting away your sunscreen and floppy hat just yet.

The hole in the ozone layer has an impact on temperatures also. Believe it or not, ozone is actually a greenhouse gas! We don't tend to call it that, because the job it does as a greenhouse gas is a very important one. When ozone intercepts harmful ultraviolet rays it transforms that energy into heat; a gap in the ozone layer means that more ultraviolet rays get through, and less heat is generated in that region of the atmosphere. In Antarctica, the ice reflects the UV light back into space (the albedo effect), so the net effect is a fall in temperature. This is part of the reason for the lack of warming in the Antarctic interior. The ozone hole and the cooling it is causing may also be helping to drive the greater winds over the Southern Ocean, because it increases the temperature difference between the equator and pole.

Most other indicators, like sea ice and glacier advance or retreat follow the same basic pattern as temperature change. The most dramatic changes are on the Antarctic Peninsula: ice shelves are collapsing and the sea ice around it is melting more quickly when spring time comes around. This is probably due to the warming waters of the ACC flowing around Antarctica and hitting the Peninsula squarely between the eyes. This warm water melts the floating ice from below. The *coup de grace* for an ice shelf is delivered by warmer air temperatures, which creates ponds of melted water on the surface. These 'meltwaters' flow into cracks and because it is heavier than ice, it forces the crack wider. The water may also refreeze and like those beer bottles we have left accidentally in the freezer, the contents expand and make the crack wider still. This process weakens the shelf until it fails in a catastrophic manner as happened in 2002 to the Larsen B ice shelf on the east coast of the Antarctic Peninsula (see Image #11). This acts a bit like pulling a cork on a champagne bottle, which releases the pent-up pressure behind it and allows the warmer ocean waters to lap at the edge of the glaciers. As a result, the glaciers that were previously held back by the ice shelf are starting to flow into the sea more quickly, which means they are shrinking – some 87% of the glaciers are in retreat on the Antarctic Peninsula.

The other major area of continental ice loss (ie glaciers on land, not the frozen sea) is the Amundsen-Bellingshausen Sea (see Image #1), the part of West Antarctica where glaciers are not protected by the Ross Ice Shelf. In other words, there is no cork in the bottle to protect glaciers in this region. The warmer ocean water is able to eat away at the ice, which you will remember is sitting below sea level in this part of Antarctica. Antarctic scientists have measured increased flow from the Pine Island and Thwaites glaciers, which aren't protected by an ice shelf like that in the Ross Sea, and could ultimately contribute 1.5 metres to global sea level. Some scientists worry that these could be the 'weak underbelly' of the West Antarctic Ice Sheet, and if they collapse, the rest will soon follow. A massive block of ice currently breaking off from Pine Island Glacier is exacerbating these fears.

Elsewhere, results are mixed. In parts of East Antarctica, the interior of the ice sheet appears to be getting a bit thicker (depending on how you measure it), although where the ice meets the ocean, there is ice loss due to the warmer waters. Furthermore, the Ross and Weddell Seas are actually increasing the extent of their sea ice (at least in area: we don't know about thickness). Of course, this information has been deployed to great effect by some climate change deniers, particularly as an argument to deflect attention from the melt in the Arctic. Alas, a bit more sea ice does not an ice age make. There are reasons why these changes are happening.

The irony of increased ice in parts of East Antarctica is that they are probably caused by warming. Remember that Antarctica is a desert, with very little precipitation falling because it is so cold. That's right – too cold for snow: go figure. Cold air is too dry to carry much moisture. As things warm up, parts of East Antarctica seem to be getting more snow. As it is not yet warm enough for the snow to melt, the amount of ice there is actually increasing. Remember in this part of Antarctica the ice is safe up on land, so the ocean can't get to it except at the fringes.

Antarctic sea ice has been increasing by an average of 1% per decade since observations began, with most of that in the Ross and Weddell Seas (the Ross Sea gaining about 4.5% more sea ice per decade). This increase in sea ice is probably due to many factors – the main one being that the fresher surface waters around Antarctica are preventing the upwelling of warmer water which helps melt the sea ice.[57]

Overall, the tale of climate change in the Antarctic may not be as stark as that in the Arctic, but that is because rising temperatures have been balanced by other forces. However, we can see in the Antarctic Peninsula and Amundsen-Bellingshausen Sea that when warming comes, its impacts are quick and dramatic. Overall, Antarctica is losing about the same amount of ice as Greenland – an average of 500 million tonnes (that is more than half a cubic kilometre) of ice each day over the last few years. This loss appears to be accelerating. This is enough to raise the sea level by around 1 millimetre each year. Most of this is coming from West Antarctica and the Antarctic Peninsula – the two hot spots.

This melting of ice is contributing to another trend that we discussed above: the freshening of the water in the Southern Ocean (in other words it is getting less salty). More rain and snowfall and reduced ice formation may also be playing a role. This is a worrying trend; remember that cold, salty, heavy water left behind when ice forms is what causes water to sink and become Antarctic Bottom Water. There are indications of reduced rates of sinking but it is early days as the changes are minute and long term observations are scarce, but the pattern fits. Oxygen levels of the water near the ice shelf are also decreasing, which also points to reduced circulation of the water. We will return to these issues later.[58]

Predictions Based on Current Trends

"Prediction is very difficult, especially about the future" – *Niels Bohr*

What do all these current trends suggest about the possible future? It is still early days, so it is hard to be sure, but taking what we do know and plugging it into our climate models produces some fairly stark results.

Firstly, it appears the cooling bonus Antarctica received from our past devastation of the ozone layer is not going to last. As ozone returns and incoming UV light once again gets transferred into heat, this could make a difference in the atmospheric temperatures that we see over Antarctica. This means we are likely to finally see increases in air temperature of around 0.34°C per decade – around 3°C over the whole century.[59]

This is a very rapid rise in temperature, but is not unheard of – we have already witnessed similar changes in the Arctic. This is a phenomenon known as 'polar amplification'; the tendency for warming to be larger at the poles. This is partly due to the fact that during summer the poles get constant sun, as well as the loss of the albedo effect. Remember if there is less sea ice around Antarctica, less light gets reflected back into space and the ocean absorbs more heat.

However, as we pointed out before that sort of temperature increase is unlikely to have much impact on the ice of Antarctica, especially the East Antarctic Ice Sheet, which mostly sits above the ocean. Even a 3°C rise won't get average air temperatures above zero, so the ice is unlikely to melt because of the warmer air, except at the edges where it touches the warming ocean. In fact as we have seen some parts of the ice sheet inland may initially get thicker under climate change.

The greater threat to the ice sheets of Antarctica are likely to come from the ocean, but it is here that things get a bit less certain. It all depends on what happens with the ACC. As we have seen, winds over the Southern Ocean are becoming stronger and moving south. It is unclear how the ACC will respond, but it seems to be becoming faster or more turbulent (or both) and moving slightly to the south.

Will this wind continue to increase and move south? Remember – this depends on the temperature difference between the equator and pole. This one is difficult to pick because on one hand carbon emissions are warming the equator faster than the pole, leading to a bigger temperature difference and more wind and current. On the other hand as the ozone layer repairs the pole will warm, reducing the temperature difference and easing the wind. We can safely say that an increase in wind is likely eventually, but in the short term as the ozone layer repairs we don't know what will happen.

We do know that the ACC will keep warming, and keep slamming into the Antarctic Peninsula, so the rapid melting of ice we have seen there is likely to continue. The rest of Antarctica has been protected by the ACC swirling around it, acting as an insulator. Along most of the Antarctic coast the ACC doesn't quite connect with the edge of the ice, in fact the upwelling of deeper cold water from the North Atlantic that takes place south of the ACC helps to insulate Antarctica from temperature rises in the rest of the ocean. As the stronger winds make the ACC faster and more turbulent this insulating effect could increase.

Longer term this may not be enough to prevent the melting of ice, due to two factors. The ACC has vast gyres – massive swirling eddies that lie between the eastward flowing ACC and the westward flowing current at the Antarctic coast. The main ones are the Ross and Weddell gyres plus a newly identified feature which we will call the Wilkes gyre. These swirling bodies of water allow some warm water to sneak through the ACC to the Antarctic coast. Also as the winds over the Southern Ocean shift south, the ACC may move south also, although this is limited to some degree by underwater topography. If the ACC starts hitting the coast of Antarctica that could spell trouble for the ice, particularly in the West where it sits below sea level.

What about sea ice? This has been confounding climate change by increasing slightly, but models predict that with the return of the ozone and continual warming of the ocean around Antarctica we will finally see the dwindling of Antarctic sea ice. While the impact will not be as dramatic as in the Arctic, the estimates are that sea ice will start reducing quite rapidly later this century, by between one quarter and a third by 2100.[60] This loss will mainly be in winter, so we will see less large fluctuations in the freeze/thaw cycle.

This loss of sea ice is likely to have feedback effects similar to what we are seeing in the Arctic. The decreased albedo from less sea ice is expected to increase temperatures by around 0.5°C per decade in the ocean off East Antarctica. Remember this is because the ice is no longer able to reflect sunlight into space. However, there could be countervailing effects too, also with an albedo impact. As the sea ice retreats, we will get more plankton growing in the open ocean left behind. As we saw in the first section, the plankton create *di-methyl sulphide* (DMS), which could stimulate increased cloud cover. As we saw in *Poles Apart* increased cloud cover could help counter the loss of sea ice by reflecting sunlight, or the increased water vapour could act as a greenhouse gas. We expect the latter effect to be the greater but we just don't know enough about clouds to tell how much greater.

The increased melt of ice, along with the reduced freezing of sea ice, and increased precipitation over parts of Antarctica will continue to contribute to the freshening of surface water. This could potentially lead to less of the dense super-cold, and salty water sinking to the bottom of the ocean which, as we have seen, helps drive the mysterious deep *thermohaline* currents.

This brings us to one of the trickiest questions: what impact is climate change having and likely to have on ocean mixing? As we have seen, this is influenced by the dynamic duo of the ACC in the Southern Ocean and ice formation in Antarctica. Ocean mixing is vital for the absorption of heat and carbon dioxide (to reduce the impacts of our greenhouse gas emissions) and for circulating nutrients through the ocean.

On one hand, the winds over the ACC are picking up, which may mean the ACC gets faster or more turbulent (or both). This could lead to more ocean mixing. On the other hand, the increased melting of ice around Antarctica may reduce the formation of deep currents, leading to less ocean mixing. In short, we don't really know.

There are also signs that the Southern Ocean is absorbing carbon dioxide more slowly than in the past, and as we have covered this may be due to the fact that warmer waters cannot absorb as much gas. This is one of the possible 'positive feedback loops' from global warming – a reason why Nature might not continue to 'correct' the impact of climate change as we'd all like to hope, and changes look set to accelerate. On the other hand this reduction in CO_2 uptake may not happen soon enough to save the ocean from increased acidification. The acidification of the Southern Ocean is expected to proceed apace, resulting in a lack of aragonite saturation by 2100. We will explore the impacts of all these changes on the critters of Antarctica in the next section.

This brings us, finally, to the vexed issue of sea level. Of all the impacts of climate change, sea level rise is one of the slowest to eventuate, but even small increases can have a massive impact, because so much of our civilisation is built around the coast. We are already seeing the knee-jerk reaction from coastal property owners, who are destroying our beaches by erecting barriers and seawalls in a Canute-ish attempt to protect their precious real estate.

As we have seen sea level rise comes partly from the increased volume of the warmer water, and partly from increased melt of ice adding water to the ocean. The melt of ice is becoming more important over time as the loss from ice sheets in West Antarctica and Greenland accelerates. The previous IPCC estimate for sea level rise was between 18-60 cm over this century, without including any dramatic acceleration in ice sheet loss, and currently the melt in Antarctica is tracking at the lower end of the IPCC estimates. However, overall sea level rises have been above trend, so the more recent estimates of sea level rise are creeping higher, within a range of 30-180cm,[61] with one metre shaping up as the best guess.[62]

Currently most of the East Antarctic ice sheet looks pretty safe; apart from a series of ice shelves and tongues, much is above the warming sea water, and air temperature rises alone are unlikely to have much impact. This is just as well, because if the whole lot melted we would be looking at sea level rises of some 60 metres.

The real jokers in the pack for sea level rise are the West Antarctic ice sheet (WAIS) and Greenland ice sheet. As we saw in the first section, the West Antarctic is actually an archipelago rather than an ice sheet sitting on land as we once thought it was. This means that the warmer waters of the Southern Ocean are lapping up against it, causing the edges to melt at an increasing rate. The WAIS makes up about a fifth of the area of Antarctica, and if the whole ice sheet collapsed, sea levels could rise by 3-5 metres. This seems unlikely to happen this century, but sudden losses in ice sheets have happened in the past and might be possible again. Could the WAIS melt? And if so, how quickly? This is clearly an important question, and lots of smart people are digging into it — literally digging, because the best answers to the question lie in the past.

What Can the Past Tell Us?

Antarctica and the Southern Ocean provide a kind of natural laboratory for science. These places are relatively untouched by human activity compared to the rest of the world. One obvious reason is that elsewhere humankind has already changed the landscape, hunted the animals and built roads or buildings. Just as archaeologists in other parts of the world can find out lots about the history of humans by digging in the ground, in Antarctica we can find out a lot about the history of the world by digging in the ice on land and sediments beneath the ocean.

There are times in the past where global conditions were similar to those we expect to be experiencing in the near future. Studying those periods helps us answer the riddles of what will happen to the WAIS and global sea level. First, we need to know what bits of the past have most in common with those we face in the future?

How the Earth Has Changed

As we saw in the first section, the world has been very different in the past. Between 145 million years ago (mya) and 65 mya the world witnessed the Cretaceous Period and the end of the dinosaur era. During this period the atmospheric concentrations of CO_2 were well above 1000ppm, and average surface temperatures were some 6-7°C warmer across the planet than they are now. There was no ice sheet on Antarctica during this time. Around 35 million years ago, plate tectonics was driving the Australian continent northwards and had opened the Drake Passage (the gap between South America and the Antarctic Peninsula), The opening of a seaway around Antarctica allowed the creation of the ACC. Over time, CO_2 levels and temperatures fell, and some ice started to appear on Antarctica. The box on the next page provides a short history of plate tectonics and its Kiwi/ Antarctic connections.

Peter's Scientific Jigsaw

Until the mid-1960s, the theory of continental drift and plate tectonics was in its infancy. In fact, it was simply a burgeoning scientific hypothesis. And it was a controversial one at that. Much of the evidence used to support the hypothesis was based on the observation that some continents seemed to fit together nicely in a kind of global jigsaw puzzle. Then a set of remarkable observations were made by scientists who were collecting magnetic data from rocks on Earth's ocean floor. These scientists knew that the Earth's magnetic field changes through geologic time and that the magnetic field at any point in time is preserved in volcanic rocks as they cool. The scientists also knew that Earth's ocean crust is made of a volcanic rock called basalt. When the scientists collected the magnetic data they noticed that the magnetic signal preserved in the sea-floor rocks produced a mirror image across submarine ridges. These mirror-image magnetic properties, which are often shown on maps as black and white stripes, showed that new seafloor was generated at the ridge and slowly flowed away from either side of the ridge and that changes in Earth's magnetic field were recorded in the new volcanic rock to produce the magnetic 'striping'. Suddenly there was a mechanism – sea floor spreading – to make the continents drift. And in a 1963 paper to *Nature*, Matthews and Vine set the theory of continental drift in motion.

Happily, it was a Kiwi scientist who helped confirm that drift had occurred. As a PhD student in the 1960s, our very own Peter Barrett discovered the first remains of a land animal on Antarctica. It was the jawbone of lizard preserved in a 220 million-year-old rock layer exposed in the Transantarctic Mountains. This not only showed that Antarctica's climate was once much warmer, but that Antarctica shared land animals with other distant continents. This suggested that the land masses must have once been joined — supporting the then controversial hypothesis of continental drift and helping to establish plate tectonic theory.

Since then, Peter's curiosity and passion for Antarctica's geological past has led him from studying strata in the Transantarctic Mountains to drilling off the Antarctic coast for climate and ice sheet history (including working with the ANDRILL programme, which we'll cover shortly). He has led over 20 expeditions to Antarctica and has been recognised for his work by geologists around the globe.

The ice sheet advanced over East Antarctica, and although it has grown and shrunk, it has not completely disappeared since then. We know this because students studying rocks in the Trans-Antarctic Mountains stumbled across freeze-dried lake beds complete with mosses, twigs and pollen, which dated from 14 million years ago. They paint a picture of an alpine landscape similar to New Zealand and Patagonia, a mountain lake dammed off by a glacier and surrounded by beech forests. Back then, these lakes were some 20°C warmer than they are today, and they are a far cry from lakes in Antarctica today, where only microbes can eke out a living.

In the last million years, average global temperatures have ranged from 10°C (with CO_2 at 180ppm) to 15°C (with CO_2 at 280ppm). This is mostly due to regular changes in the Earth's orbit around the Sun which cause ice ages and intervening warm periods – the so called glacial and interglacial cycles where the reach of polar ice sheets have pulsed (even pushing permanent ice sheets over northern Europe). We are currently in an interglacial interval. Temperatures reached the natural peak warmth of the current interglacial period about 9,000 to 5,000 years ago, and since then temperatures have generally declined, arguably on the slow progression into the next ice age. Of course, we have lately reversed this trend back toward warming, courtesy of our carbon emissions.

In Antarctica, temperature variation between glacial and interglacial periods has been much larger at up to 9°C as per the ice core record. There's that albedo effect again, playing havoc with the polar climate by raising temperatures when the ice melts. Coral reef records suggest that during warmer periods over the last 400,000 years, sea levels have been 4-6 metres higher than today, with temperatures 2-5°C warmer. Conversely in glacials, or ice ages, sea level has plunged 120m to 130m below present as ocean water was taken up to grow the ice caps (mostly in the Northern Hemisphere). Think of it: with sea level that low, you could walk from Farewell Spit to Cape Taranaki, or stroll across the Bering land bridge and populate the Americas, as happened just 20,000-40,000 years ago.

So throughout the Earth's history, there are examples of many different climates. If we can look back through the past, maybe we can see what happened in those different climates to get some clue to our own future. Scientists have found a way to do this in Antarctica.

Within the ice of Antarctica are trapped tiny air bubbles, each of which acts as a snapshot of the planet's atmosphere through time. From these air bubbles, scientists can extrapolate a rich picture of the planet that existed at the time. And the beauty of these measurements is that they aren't proxy measures: once released, the gas in these bubbles is the air that was wafting about in previous ages of the world. The deeper you dig, the further back you go in the atmospheric record. This is why, since the late 90s, scientists (including Kiwis) have been drilling ice cores in Antarctica. These have yielded a lot of information about the atmosphere over the past million years or so.

For example, the air bubbles provide an understanding of both the atmosphere as well as climatic conditions that existed in the past. This has helped us understand the natural cycles that the planet goes through, and even helped to confirm the links between greenhouse gases and atmospheric temperatures. Without this crucial information, we would still be squabbling over whether there was man-made climate change at all. For example, these ice cores have told us what happened last time the planet warmed. As the sea ice retreated in the Southern Ocean, the warmer ocean waters were able to release the gases they were retaining, including CO_2, which seems to vindicate the fear that de-gassing from the oceans will form a positive feedback (just because it's positive doesn't mean it's good!).

So can the past tell us what is likely to happen to the West Antarctic Ice Sheet (WAIS) if temperatures continue to rise as atmospheric CO_2 levels soar past 400 ppm? In the last million years, the climate has generally been colder or the same as it is now, and CO_2 levels have stayed within the tight range of 190 to 280 ppm. The last time CO_2 exceeded 400 ppm was around 3-5 million years ago when average global temperatures were 2-3°C warmer than today and sea level was (at different times) 10-30 metres higher. As ice cores don't go back that far in time, scientists needed a new approach to unlock the secrets of the WAIS.

ANDRILL and the WAIS

To examine warm periods that predate the reach of the oldest ice core, we need to study rocks and sediments instead of ice cores. This is what the ANDRILL programme set out to do, again in Antarctica. What the heck is ANDRILL? All is revealed in the box below.

ANDRILL[63]

Over two Antarctic field seasons during the summers of 2006 and 2007, a large team of researchers, drillers, educators and support staff from four nations met in Antarctica to recover and study two 1 km-long sediment cores pulled from the sea-floor beneath the Antarctic sea ice and an ice shelf. These two drilling projects were completed as part of the ANDRILL program (ANtarctic geological DRILLing), a special international collaboration, involving well over 100 people from the United States, New Zealand, Italy and Germany who worked together to drill deeper than ever before into the Antarctic continental margin. ANDRILL's drilling program was managed by Antarctica New Zealand and included scientists from Victoria University, Otago University, Canterbury University, and GNS Science. These were a couple of the biggest projects undertaken as part of the International Polar Year.

The aim of the ANDRILL Program is to recover sediment cores from beneath the Antarctic seafloor in order to unlock the secrets of the geological past. The two inaugural projects aimed to use cores and the data they contained to help decipher Antarctica's climatic, glacial and tectonic history over the past 20 million years. In particular, the scientific team on ANDRILL's McMurdo Ice Shelf Project wanted to uncover the history of the Ross Ice Shelf and West Antarctica, with a particular focus on the last five million years. As we mentioned above, there is a theory that the Ross Ice Shelf (RIS) functions as a cork in the champagne bottle that is the West Antarctic Ice Sheet (WAIS). Remove this cork and the edges of the ice sheet are suddenly bared to the warming ocean, causing melting to increase and ice streams (flow) to accelerate until much of the ice sheet disappears. Sediment cores from the Ross Sea can tell us how the RIS and WAIS have ebbed and flowed in the past.

The drill team on the McMurdo Ice Shelf Project worked at a remote camp located 11 km east of Scott Base in the Western Ross Sea. They managed to recover a mammoth 1.2 km long rock and sediment core from beneath the seafloor. But before they were able to get their drill string to the seafloor, they had to melt a hole through 85 m of slow-moving ice and then lower the drill bit through 850 m of seawater to the ocean floor. It was a real challenge to melt and maintain a hole through the ice so that the drill system could operate for the month it took to drill the 1 km. The drill team also had to deal with the rising and falling tides: amazingly, these still happen beneath the floating ice shelf. Another challenge was to develop a drill that was sensitive enough not to destroy soft sediments and that was nevertheless still capable of drilling through hard rock. The logistical problems this project posed led to many innovations in sub-ice drilling and coring technology.

The core they extracted was the longest and most complete from Antarctica. When they got it back to the amazing lab at McMurdo Station (US), it was sampled and examined by teams from different countries – it's a kind of friendly United Nations land grab. The core contains layers of sediment that have been deposited as the West Antarctic ice sheet has grown and retreated across the area over the past 14 million years.

Sediments underneath the ice or the seabed can tell scientists a lot, but it takes a bit of forensic detective work. Scientists have to make estimates on conditions and climate that were prevalent at the time based on the rock, mud and life forms they find in the core. For example, when shells are formed they incorporate magnesium which can, through various complicated and ingenious means, be used to identify the temperature at the time of shell formation; elements such as barium and zinc help them to ascertain the production of plankton; the presence of stable isotopes of oxygen provide insights into the volume of ice stored in the polar regions as well as temperatures. The sophistication is improving all the time as new analytical technologies find their way into the lab. Similarly sediments under the Southern Ocean can be used to deduce information about the ocean temperatures, salinity, circulation, the presence or absence of ice, and how much life there was at various points through time. The evidence is there, but not directly, so it is a bit like piecing together the plot in a detective story.

Sure enough, scientists found at least 60 glacial-interglacial cycles in the sediment core, which reflect climate fluctuations (cold and warm periods), and the growth and collapse of ice sheets and shelves. These cycles of ice sheet advance and collapse were mostly due to predictable changes in the Earth's orbit that happen every 21,000, 40,000 and 100,000 years.

A second sediment core recovered during the Southern McMurdo Sound Project managed to extend the geological record of ice sheet behaviour even further back in time – to around 20 million years. This interval includes a period of time when Earth was significantly warmer than present and atmospheric CO_2 levels were estimated to have been as high as 600 ppm. The insights from this work will be invaluable in helping us to understand and predict the changes that are underway as the polar regions warm in response to our emission of carbon dioxide.

Results from ANDRILL's projects are helping to build a more complete picture of the world's climate system and its influence on Antarctic ice sheets. They align well with other drilling records, and geological records of sea level rise from around the world.

As we discussed in the Race for Resources section, science in Antarctica brings many other benefits, and ANDRILL provides a good example of this. In particular, it helped maintain and enhance collaboration between the United States and New Zealand and provided a platform for diplomatic interaction. New Zealand contributed some scientific grunt to the project (we have some of the world's best polar climate scientists) as well as cutting-edge and sophisticated drilling technology and project management. Just after the ANDRILL Projects were completed, New Zealand was offered the chance to lead a Ministerial discussion between countries involved in the Antarctic and Arctic about the successes of the International Polar Year (of which ANDRILL was part). Our Foreign Minister Murray McCully even obtained his first full bilateral meeting with Hillary Clinton early on in the Obama Administration's first term. Clinton cited the discoveries of Antarctic geological drilling among the major achievements of the IPY. Was this series of diplomatic achievements associated with ANDRILL, or a mere coincidence? Well, there doesn't seem to be much doubt that a good way to cosy up to the Yanks is to get stuff done with them.

What have we learned from this latest journey into the past? We know that for much of the past several million years the West Antarctic Ice Sheet has advanced and retreated in response to regular, natural variations in Earth's climate that are mostly driven by changes in the amount of solar energy reaching Earth's surface. ANDRILL's scientists can tell this because they found packages of rock that include three kinds of sedimentary rock type: diamictite, sandstone, and diatomite. These rock types reflect three environmental 'phases' at the drill site. The first phase is represented by sedimentary rock (diamictite) that was deposited beneath thick ice sitting on the sea floor and records intervals of time when the whole Ross Sea region was covered by a thick sheet of ice. The second phase is represented by sandstone and mudstone, which were deposited beneath floating ice shelves and records periods similar to the present when the West Antarctic Ice Sheet did not extend across the

Ross Sea. The third phase is represented by sediments that are composed almost entirely of marine algae called diatoms. These sediments record times when the Ross Sea was all open ocean and when the largely marine-based West Antarctic Ice Sheet was likely gone. Each of these packages of sedimentary rock have been dated and span regular time intervals that match the length of the good old Milankovic cycles. Therefore scientists infer that the natural variability in solar input due to orbital changes drove the advance and retreat of the West Antarctic Ice Sheet for much of the past several million years.

But we also know that other (non-orbital) factors can affect the amount of solar energy that is trapped at Earth's surface. By studying rocks of early Pliocene age in the ANDRILL cores, scientists were able to determine what happened in Antarctica when atmospheric CO_2 levels were only slightly higher than present and at levels we will reach in the next several decades. When scientists recovered the early Pliocene rocks they were amazed at what they saw. Instead of the regular cycles of diamictite, sandstone, and diatomite that were typical of younger part of the core, there was a long 80 metre-thick section composed entirely of diatomite. This was a remarkable discovery – the Ross Sea was once ice free for some three to five hundred thousand years, and remained ice free through several orbital cycles. During this extended interval of time, the Ross Ice Shelf and West Antarctic Ice Sheet collapsed and did not grow back. Perhaps the increased levels of CO_2 caused temperatures to remain high enough to restrict ice sheet growth even through 'background' changes in orbital factors.

Geological data from several locations around the world, including New Zealand's Wanganui Basin, indicate that sea-level was between 10 and 30 meters higher than present during the early Pliocene. We know from the work done by ANDRILL's scientists that West Antarctic ice sheet collapse contributed at least 3.5 m of the sea level rise. The Greenland Ice Sheet also likely melted at this time and would have contributed up to 7 m to the sea-level increase. Thermal expansion of the oceans can probably account for another 2 m or so.

Meanwhile, most geological evidence suggests that the East Antarctic Ice Sheet remained largely intact at this time because most of it sits on land. However, geological data from cores and outcrop around the continent suggest that ice was lost from those edges that were in contact with the ocean surface, which was up to 4°C warmer than today.

So geological records and sediment cores provide a way to examine the riddle that looking at modern trends couldn't answer, simply because our modern records of change are so short. In short, under conditions where the CO_2 level was similar to that prevailing today, we know that the polar regions were quite different, even under the coldest orbits. With CO_2 levels at very similar levels to today, temperatures eventually rose enough to melt the WAIS. This probably added some 3.5m to the global sea level, and total sea level was probably 10-30 m above today after contributions from Greenland, Arctic Islands and other land-based ice.

What Does All That Mean for New Zealand?

We will finish this section by taking a quick look at what these trends mean for us in New Zealand. There will be winners and losers from a changing climate, and it seems that New Zealand overall could relatively be a winner. In a world of increasing drought and flood, our climate seems likely to stay relatively benign, and could even allow us to grow more food, although we are likely to be subject to more weather extremes. Furthermore we will not escape rising sea level.

As mentioned, the weather trend under climate change seems to be pushing the westerly winds south below New Zealand and over the Southern Ocean. This leaves us with less unsettled weather and could mean less rain and warmer temperatures for the west coast of the South Island. Yay! Well, not quite. A lot depends on how these Southern Ocean climate patterns interact with those we get from the north (El Niño and La Niña). These normal climate fluctuations are likely to become more intense, as we are experiencing with the present 'La Niña' phase.

As we have seen, the ocean also has a huge impact on the world's weather and climate. There is emerging evidence that some extreme weather events (e.g. droughts, heat-waves and big downpours of rain) are happening more frequently in some parts of the world as a result of climate change. The economic cost of climate-related disasters has certainly increased,[64] resulting in more insurance claims and higher premiums.[65] As shown by the graph below, the number of weather related disasters with insurance implications has risen steadily since 1980.

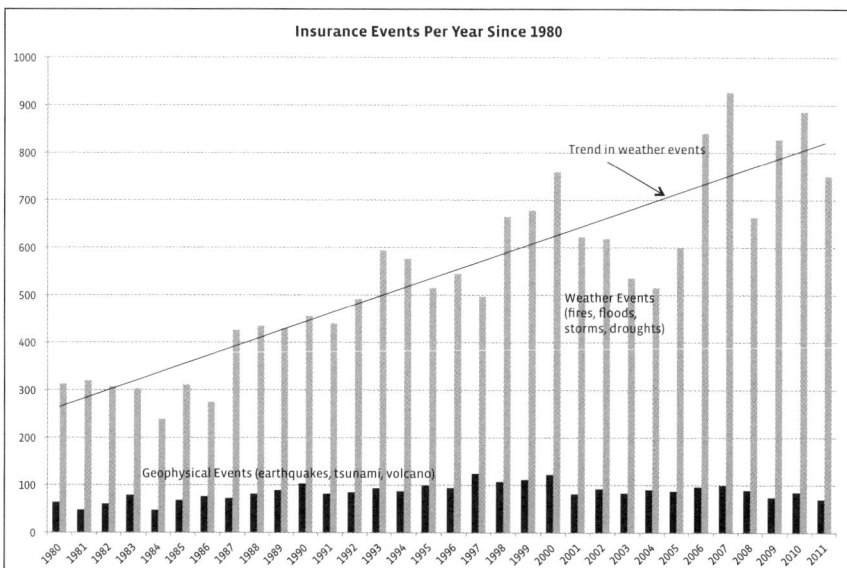

Insurance Events Per Year Since 1980

Source: © 2012 Münchener Rückversicherungs-Gesellschaft, Geo Risks Research, NatCatSERVICE

The increased number of extreme weather events around the world is combining with rising sea levels to increase flooding risk – estimates are that the UK will face 30 times the number of floods by 2080.[66] In the case of coastal cities like New York, this future carries a high potential price tag – severe storms could cost an estimated $100-$200 b because of the impact on transport and infrastructure.[67] The flooding in Queensland over the 2010/2011 summer cost $A30 billion, a disaster that was almost repeated in 2012. 2011 was a record breaking year, not just because of the Japan and Christchurch earthquakes – as bad as they were – but also because of weather and climate caused disasters.

This brings us to sea level rise, one of the more concerning aspects of climate change. The rate of sea level rise is now over 3.1 mm per year, which suggests a 30 cm rise over the century. Remember that we are now seeing predictions of a 1 m rise by 2100, and the top end of the estimate range could reach two metres. About half of this is expected to come from the WAIS and Greenland ice sheet.

However, this is only a short-term view. We have little prospect of limiting the atmospheric concentration of CO_2 to below 400 ppm, so we have already locked in a temperature increase of 2-3°C. Over time, this will continue to heat the ocean, thereby expanding it and melting more ice to boot. The WAIS melted last time temperatures rose by that amount, so marked rises in sea level seem inevitable. The only good news is that this will probably take 400-700 years to fully play out.

A sea rise of one or two metres is likely to flood very low lying areas, like river deltas and wetlands. These are important environments and will be a major loss to our ecosystem, but in New Zealand not many people live in these areas. The countries hit first will be the likes of Bangladesh, Myanmar or the Pacific Islands. Even in the United States, researchers estimate that 3.7

million people could be forced to move with a sea level rise of only one metre.[68] For most of us in New Zealand, however, our precious beachfront property is probably safe for this century, except perhaps from storm surges during high tides. One of the biggest impacts on our country this century could be the rising numbers of climate change refugees as people from Asia and the Pacific Islands are forced to leave their homes as they sink beneath the waves.

The plight of our Pacific neighbours seems particularly bleak. Sea-level rise, storms, overpopulation and the degradation of coral reefs will threaten the livelihoods of anyone living a stone's throw from the coast. Towns and key infrastructure are all at risk, and the risk will rise as coral reefs die off and beaches erode. By 2050, water resources in many small islands could be pushed to the point where they cannot meet demand during low-rainfall periods. In the extreme case, islands will probably end up submerged, a fate which seems unavoidable for Tuvalu and Kiribati given the international dilly-dallying over carbon emissions.[69] The twin threats of population growth and climate change have already impinged upon the water supplies of major Australian cities which are in various stages of getting energy-hungry desalination plants online. Let's hope the energy for this comes from renewable sources, because most electricity in Australia currently comes from coal plants, which emit CO_2, which lead to warmer climates that exacerbate water and energy shortages...[70]

Regrettably, there will be plenty for future generations to deal with as a result of our actions today. Remember that if the WAIS melts, as it did the last time CO_2 reached current levels, sea levels could rise by 3-5 metres. Five metres may not sound like much, but it is enough to devastate many very expensive beachfront properties. Even five metres is enough to wipe out the Auckland Wellington and Hutt waterfronts (and airports!) and we would be kissing poor old Invercargill and Napier goodbye. Rural areas are in far bigger trouble, as five metres is enough to threaten low-lying areas like the Hauraki Plains, Blenheim, Kaiapoi and Foxton. Lake Wairarapa would no longer be a lake at all, but a new shallow harbour. Maybe time to move to Hamil-tron. It's clearly the place to be, with the lowest risk of any natural disaster anywhere in New Zealand.

Summary

Our Far South, and particularly Antarctica, is a vital place for understanding climate change. Its relatively untouched, icy environment provides the perfect laboratory for researching how the climate has changed in the past, and what those changes have meant for the world. Kiwis are heavily involved in the incredible task of piecing together this history from ice and sediment cores.

We also need to closely monitor how climate change is affecting *Our Far South*. As in the Arctic, global warming is likely to impact more heavily on Antarctica than elsewhere. Because of the loss of the albedo effect from the ice, the polar regions are warming more rapidly than the rest of the planet, making them a barometer for change. However, thanks in part to the hole in the ozone, this impact is not fully evident in Antarctica.

Finally, *Our Far South* is important because of its influence on our climate and ocean. Antarctica and the Southern Ocean are major drivers of the world's climate and ocean. Changes down there will drive changes in our sea level, climate and in the fish we pull up at the end of our line.

What can we expect from climate change in the future? In short, we will start to see more dramatic changes in Antarctica as the ozone repairs itself over time and CO_2 continues to build up in the atmosphere. We can expect the sea ice to retreat over time — by up to 30% by 2100. The upper layers of the Southern

Ocean will continue to become warmer and fresher. Stormy weather will move from over New Zealand to over the Southern Ocean, so the ACC will become faster and/or more turbulent as a result. The implications for ocean mixing are unclear: this needs close monitoring.

At current or rising levels of CO_2, the ice sheets of Antarctica will also shrink over time, eventually (it seems) resulting in the collapse of the West Antarctic Ice Sheet. Thanks to expanding sea water and melting ice, sea levels are expected to rise by about a metre this century, but the eventual total could be more like 10-30 metres unless CO_2 levels in the atmosphere are brought down. Meanwhile the ocean will be quietly becoming more acidic as long as it continues to absorb CO_2 from the atmosphere.

All of these changes, combined with the impacts of humankind's commercial exploitation could potentially have a huge impact on the wildlife of *Our Far South*. This has implications for the species that we are trying to save down there. The next chapter will look at the threats that both the race for resources and climate change are having on the conservation effort in *Our Far South*.

CONSERVATION

As we have seen, much of humanity spent the nineteenth and most of the twentieth centuries locked in a race for resources that devastated much of the wildlife in *Our Far South*. Whether it was through direct exploitation, or through the indirect impacts of fishing, habitat change and the introduction of new predators, many species have been pushed to the wall. Since then, we have been busily trying to repair this damage and making sure we don't repeat these mistakes elsewhere (such as with toothfish).

Why is conserving *Our Far South* so important? What species are at risk, what are the greatest risks they face, and what can we do about those risks? We will start by looking at the ecosystems of *Our Far South* — but first we need to understand why we're worried about species becoming extinct at all.

Every time we lose a species, the whole ecosystem becomes more unstable. The number of species that exist in an ecosystem is known as *biodiversity,* and having lots of different species in an ecosystem is important for similar reasons it's important to spread your financial investments across many different kinds of assets. If the housing market tanks, for example, then the rest of your portfolio can keep you from going under. Same with the ecosystem – if something bad happens then there is a greater chance the ecosystem won't collapse if there are plenty of species around.[71] For example, a blight that affects one species won't spread so voraciously if that species is interspersed amongst many others. If it's a monoculture, however, it is more likely to wipe out the whole species. There is diversity within species, too

– just look at your fellow humans – and this also seems to help. Our genetic diversity stopped humans getting completely wiped out by the bubonic plague back in the Middle Ages.

Biologists compare the stable state of ecosystems to a boulder resting on a valley floor. Delivering a shock to an ecosystem is like rolling the boulder up the valley wall: let it go at a suitable point, and it will roll back down. This resilience to shocks is all the more important in today's world, where the big shocks include what humans are doing to the climate and the landscape or seascape as our race for resources rumbles on unabated.

Sadly, biodiversity is decreasing in most parts of the world – the latest WWF global "Living Planet Index" of biodiversity shows a 30% decline since 1970, particularly in the tropics.[72] What happens when an environment has had more change than it can handle – when you roll the boulder to the top of the ridge? Sometimes it will roll down into a different valley, and the ecosystem just changes to a different way of working; but sometimes the ecosystem collapses altogether. We talked about the rise of ocean 'dead zones' – places in the ocean where most animals can no longer survive — in our book on fishing *Hook Line and Blinkers*.

In colder climes, we tend to find fewer species of plants and animals than in the tropics, although those animals that can handle the cold tend to exist in large numbers. So most of *Our Far South* is not very biodiverse (with the comparative exception of the seafloor in places like the Ross Sea). Lower biodiversity means that losing a species is a bigger deal.

Maintaining the south's biodiversity through conservation is also important because of the high degree of *endemism* in *Our Far South*. Having so many species living there that don't occur anywhere else tells us if we lose them from *Our Far South*, we lose them forever. This endemism has occurred partly because *Our Far South* is so isolated, partly because there are species there that have been killed off everywhere else thanks to human activity and partly because of the unique environmental conditions to which life has had to adapt, such as the extreme cold of Antarctica.

Finally, *Our Far South* is also very important for the *productivity* of the whole world ocean — in other words how much life the ocean can support. We've seen what a massive influence the ACC has over this because it helps mix the ocean and get the nutrients to the surface for algae to use and grow, and the area where sea ice thaws is one of the most productive parts of the ocean on the whole planet. For this reason, *Our Far South* plays a major role in determining how much life there is in the oceans – how rich and populous the entire aquatic food web is, from fish and squid all the way up to sea lions, whales, birds and even humans.

The Ecosystems of Our Far South

Like men, most animals will follow their stomachs. Most life starts with plants. While the subantarctic islands are covered in plants, they only have a piffling 70,000 hectares of land to grow on, and a lot of that is barren rock. As for the White Continent, the best it can manage is a few mossy plants on the 1% of area not covered by ice several kilometres thick. So really, life in *Our Far South* is shaped by the ocean and the plants growing in it, known as algae or *phytoplankton*.

Algae, like plants on land, introduce energy (provided by the sun) into the marine food chain. Without these little plankton factories, the whole food chain falls over. As we've seen, algae is dependent on a few key factors, namely the right mix of nutrients, trace elements and light, temperature and the influence of wind on the upper ocean. There are not too many places on the planet that get this balance just right. We can see this in the colour enhanced satellite image (Image #12), which shows where algae is most abundant. This is also generally where the animals are.

As we have seen, thanks to the Antarctic Circumpolar Current, the Southern Ocean is chock-full of many of the nutrients that plants need to grow, like nitrates. This helps make these oceans more productive. Warmer oceans, such as those to the north of Australia and surrounding many islands in the South Pacific, lack the main organic nutrients such as nitrates and phosphates. The large layer of warm water at the surface prevents nutrient-rich cold water being cycled up from the bottom of the ocean. Without these nutrients, the tropical regions are basically ocean deserts.[73]

As a result, the cooler, temperate oceans are generally more productive than the tropics. But you'll notice on Image #12 that with the exception of places like the Ross Sea on the edge of Antarctica (the area where the sea ice melts), less algae grows in the Southern Ocean than elsewhere. Given the high level of nutrients in the Southern Ocean, this has puzzled scientists since the 1930s. Of course, in winter the lack of light and the covering of sea ice are going to be a problem for algae trying to photosynthesise, but the same is true of the very productive Northern Hemisphere. This tricky little riddle became known as the 'Paradox of Antarctica'.

One factor that is different about the Southern Ocean is the ACC, and this probably makes life difficult for algae. When those westerly winds blow, the speed and power of the churning ACC is such that algae can get swept away from the surface to water 80 metres deep within the space of a couple of hours. This takes them away from the sunlight before they have time to grow. So the ACC helps ensure the Southern Ocean has a nutrient-rich, algae-poor surface layer.

Yet even this wasn't enough to explain the paradox. It has only been in the last 25 years, as measuring equipment has got more sophisticated, that scientists have noticed another, incredibly subtle but hugely important difference between the Southern Ocean and the Northern Hemisphere: the lack of iron. This element is only needed in minute amounts by algae, so a trace (and we're talking in quantities less than parts per billion) can make a huge difference.

Iron usually gets blown into the ocean in the form of dust from deserts on land. Because of the lack of land in the Southern Ocean, there is no terrestrial source of iron, whereas the Northern Hemisphere's oceans get plenty of dust blowing off deserts in North Africa and Asia. In the Southern Hemisphere, we have the Kalahari and most of Australia playing the same role, but these areas are quite far north. Only the Atacama in South America can play the same role for the Southern Ocean — hence the high productivity of the ocean in this region just north of Antarctica. Funny as it sounds, the algae in *Our Far South* are anaemic, and this limits their growth. Sounds like they need to get hold of some good Kiwi beef and lamb! But seriously, this revelation has sparked speculation among scientists that adding iron to the Southern Ocean could make a massive difference to our oceans and climate, as detailed in the box below.

Iron Fertilisation[74]

Some climate change records suggest that at times in the past, iron levels in the ocean were higher, possibly as a result of larger deserts. During this period, CO_2 in the atmosphere fell. This has led to a theory that if the Southern Ocean were fertilised with iron, it could result in massive blooms of algae. If some of this drifted to the bottom of the ocean, it might store away carbon and reduce greenhouse gas levels in the atmosphere.

Our own NIWA scientists decided to test this idea in the Southern Ocean using their research vessel *Tangaroa*. Over a 14-day experiment, they added an iron salt to the water along with a chemical that would allow them to trace the iron through the ecosystem. At one stage, they lost their test site when the Southern Ocean got up to its usual tricks and sent some ten-metre waves their way. They found the site again, but they needn't have worried because the results of the experiment could be seen from space: it was that dramatic!

The experiment resulted in a massive bloom of algae, some 1000 km² in size. The anaemic state of the Southern Oceans was proved beyond doubt. But the benefits in terms of reducing greenhouse gases in the atmosphere were not so great. The algal bloom soaked up a lot of carbon dioxide initially, but as they rotted, much of that was expelled back into the atmosphere. Some of the gases expelled were even worse greenhouse gases than carbon dioxide, like nitrous oxide. All in all, the net effect on greenhouse gases was probably neutral, so the idea of iron fertilisation as a way to stave off climate change needs a bit more work.

Of course, the other side benefit of iron fertilisation could be to increase the amount of life in the ocean in Our Far South. This might be a boon for our fishing industry, or the animals on our subantarctic islands. But as with all complex systems, it is really difficult to know what the knock-on effects of this kind of meddling might be — just like putting extra nitrogen in our rivers and lakes. Remember that the currents from the ACC spread nutrients around the world. What if iron fertilisation used up all those nutrients in the Southern Ocean, preventing algal blooms in another part of the world and thus starving food webs?

There is one part of *Our Far South* that can avoid the problems of excessive churn and lack of iron, making it a productive place, if only for a short while. If you take another look at Image# 12 you will notice that the area close to Antarctica, such as the Ross Sea, is relatively productive, albeit for a limited time only.

The melting sea ice provides all the ingredients, including iron, needed for a bloom of algae. Once the sea ice has melted, the 24-hour summer sunlight can penetrate into the surface of the ocean. The melting ice also creates a thick layer of light, fresh water on the surface. This provides some respite from the churn of the Southern Ocean, which means that algae don't get swept out of the range of sunlight. Finally, the melting ice seems to contain tiny traces of iron, perhaps picked up by glaciers as they grind up against rock, or dust blown on the wind from the few ice-free parts of Antarctica. The result is an Antarctic-style feast. This was quite visible to us on the *Our Far South* voyage, as some parts of the Ross Sea were soupy green due to plankton, whereas other parts were the deep blue of empty ocean.

Now that we understand the forces that are driving the food web in *Our Far South*, let's take a quick look at the unique ecosystems that we get as a result. First up, we'll look at the subantarctics, then turn to Antarctica.

Subantarctic Islands

The subantarctic islands have a lot of unique life crammed in a pretty small space. There isn't much land in the Southern Ocean – the six island groups we're discussing are pretty much the only ones in the whole Pacific Ocean region. So these islands are like liferafts for the animals that have to come ashore to rest, breed and raise their young. As we've noted, our subantarctic islands are also remarkable for their high level of endemism: the bird and plant life, especially the albatrosses, cormorants, land birds and 'megaherbs' are unique to the islands.

Let's start with the birds. The subantarctics are famous for the huge numbers of seabirds and penguins that use them as a place for nesting. There are 120 bird species nesting on the islands in total, including 40 seabirds, which is a quarter of all seabird species in the world. For this reason, our waters have been called the seabird capital of the world, and while we're not tops, we're certainly up there. New Zealand is definitely number one in the world for endemism: we have the highest numbers of seabirds that don't live anywhere else. This is remarkable when you compare the tiny dots that are our subantarctic islands with the seabird-infested seaboards of Chile and the United States.[75]

The king of seabirds is the albatross. These majestic creatures are built for the Southern Ocean; in fact, so big are they that they rely on the blustery winds even to get off the ground. They can lock their enormous wings in place and glide effortlessly on the westerly wind for huge distances to look for food. The subantarctic islands support major populations of 10 of the world's 22 species of albatross, of which six breed nowhere else. These six include some of the 'great' albatrosses (Antipodean, Northern Royal and Southern Royal), which earn their illustrious title because they have the largest wingspans of any bird; up to 3.5 metres for the Southern Royal (see cover image). Three of the smaller albatrosses (you'll sometimes hear them called mollymawks, an old Norwegian term meaning 'silly gull' that has stuck in this part of the world), namely the Campbell, Salvin's and white-capped, also breed only in the subantarctics.

Petrels are equally at home in the Southern Ocean. They are smaller, but unlike albatrosses, they can dive into the water rather than simply feed on the surface. Interestingly, they get their name from the ability of some petrels to hover just above the waves dangling their feet in the water, looking as though they're walking on it, just as St Peter (apparently) did. Of all the world's petrels, shearwaters, fulmars and prions, 21 species (30%) breed in our subantarctics. Around 3.5 million sooty shearwaters nest on Snares Island alone. Because of the distance between islands, many of the shag (cormorant) species found in the subantarctics are each found only on one island. There are also six penguins in the subantarctics, four of which are only found in our region (Snares crested, Royal, yellow-eyed and erect-crested).[76]

One of the world's rarest members of the sea lion family also calls Auckland and Campbell Islands its home. Around 95% of the world population of New Zealand sea lion (formerly known as Hooker's sea lion) breed in the Auckland Islands. Their total breeding population is estimated to be around 9,000. As we will see shortly, it has been in steady decline for some time. They tend to return to where they were born to breed, which is why it surprised everyone when a family recently turned up on a Dunedin beach. After being hammered by sealers in the nineteenth century, and driven off beaches by human settlement, the New Zealand sea lion is just beginning to recolonise mainland New Zealand.

Southern elephant seals are also present in *Our Far South*, breeding at Macquarie, the Antipodes and Campbell Islands. These massive creatures caused a scare with a large fall in population since the 1950s, but this seems to have been arrested. Populations of fur seals, on the other hand, have been recovering strongly since the dismal days of sealing. Auckland Island also provides a critical breeding site for the southern right whale, which is staging a slow but steady recovery since whalers almost pushed them to extinction in this part of the world by the early 1900s.

The seabirds and mammals are certainly the most numerous of the subantarctic wildlife, but they don't have the place to themselves. Less well known are the large numbers of rare and endemic land birds. Many of these are so rare because they have been isolated from the rest of the world for so long that they have evolved into new species. Some 15 species of land bird are only found on these islands, including the Campbell Island snipe (more on this amazing discovery later) and one of the world's rarest ducks – the flightless Campbell Island teal. The Antipodes Island parakeet earns its crust by hanging around penguin colonies and eating the scraps. Even the insects get points for rarity; there are endemic weta on each of the island groups, and some 40% of the insects described on Campbell Island are endemic.

Our subantarctic islands also have the most diverse range of plants of all the islands in the Southern Ocean. This is partly because they are warmer than other subantarctic island groups, sitting as they do north of the Subantarctic Front. Their isolation has helped minimise the damage from human influence. The Snares and two of the Auckland Islands (Adams and Disappointment Island – so known because it disappointed sailors, not scientists – are especially important because their vegetation hasn't been modified by human or alien species (pests). Campbell Island is also regenerating after aborted attempts at farming, and is now cleared of rats and sheep.

All-in-all, the islands have 233 plants, of which 196 are indigenous, six endemic, and 30 are rare. The Auckland Islands have the southernmost forests in our part of the world, dominated by southern rata and the southernmost tree ferns in the world. On other islands, the plant life is dominated by so-called 'mega-herbs', which along with the rata, flower in a spectacular fashion. No-one knows why the plants put on such extravagant displays – it may be to attract pollinators, absorb more sunlight or simply to keep warm. South of the Auckland Islands, it is too cold for trees to grow, and without animals that munch on plants, the remaining herbs and low-growing flowers have been able to grow to huge proportions.[77]

Merganser

Even the pursuit of science has occasionally taken its toll on the biodiversity of Our Far South. The merganser was a large, fish-eating duck that was probably once found throughout New Zealand. The arrival of humans devastated the mainland population, but the merganser lived on at the Auckland Islands. As pigs and cats got established on the islands, the merganser population was pushed to the brink.

In 1902, the Governor-General of New Zealand, the Earl of Ranfurly, went on a trip to collect specimens for British museums and the glory of the Empire. This expedition managed to shoot two specimens of the five mergansers that were spotted in the Auckland Islands. These were the last two of their kind ever seen – a right royal stuff-up. So New Zealanders are obliged to view Lord Ranfurly as something of a mixed blessing: he gave us the Log o' Wood, but played his part in depriving us of a cool and unique bird.

Antarctica

South of the ACC, the species tend to be quite unique. Part of the reason for the high levels of endemism in Antarctica is that ever since its emergence, the ACC has isolated Antarctica from the rest of the world. The creatures that live there have also had to adapt to the brutally cold environment. Even the local phytoplankton (types of diatoms) are uniquely adapted to keep photosynthesizing at the same rate as other species, despite the cold temperatures, huge variations in light, and lack of iron in the water. Sloganeering scientists have dubbed diatoms 'nature's nanotechnologists', because they grow beautiful microscopic structures out of silica (the stuff we use to make glass) that would put the most accomplished glass blower to shame. But there is perhaps no better example of unique adaptations to the local environment than those exhibited by the toothfish, as described in the box on the next page.

Antifreeze Fish

Antarctica has some of the coldest ocean waters on earth. As we saw in the first section, it is possible for temperatures there to fall *below* the freezing point of fresh water. Warm-blooded mammals like whales and seals have the ability to warm themselves up, and so with a bit of insulating blubber, they can survive in these conditions. But fish are typically cold-blooded and nearly every fish on the planet would freeze into a fish-sicle if it tried to brave Antarctic waters. This is why there are no sharks in the waters of Antarctica, making killer whales public enemy number one for any creature that values survival.

The Antarctic toothfish and its cousins of the *Nototheniidae* family, however, thrive in this icy environment. How do they do it?

Antarctic toothfish have developed a handy trick that allows them to survive in sub-freezing waters. They have a special protein in their blood that acts like antifreeze. By substituting this unique antifreeze 'glycoprotein' for haemoglobin (the iron compound that carries oxygen in the blood of most organisms), Antarctic toothfish are able to keep their blood from freezing. They can manage with less haemoglobin because the waters of the Southern Ocean are so oxygen-rich anyway. This is a remarkable evolutionary solution to the problem of surviving in the frigid waters of the Antarctic.

As a result of these innovations, the *Nototheniidae* dominate the Antarctic ecosystem, at least from a fish perspective. Over half the species in the area are from this group, and they make up 90% of the specimens collected so far.[78]

These fish are endemic to Antarctica, although one stray was found near Greenland, possibly hitching a ride on the cold, salty deep water currents. Remarkably, there are similar fish living in the Arctic, but they have a completely different antifreeze protein. This suggests that fish at both ends of the planet have evolved similar survival strategies, but completely independently.

The other defining characteristic of the Antarctic ecosystem is its boom and bust nature. As we've seen the bloom of phytoplankton and krill that occurs in the Antarctic summer means that the region can support an incredible array of life, including squid, whales, penguins, seals and seabirds. Alongside the recovering whale populations, the six species of seals in the Antarctic are the biggest consumers of krill. By number, Antarctic seals make up 60% of the world's total seal population: in fact the crabeater seal is the most populous large mammal in the world (apart from humans, of course!) with a population of 10-15 million.

Some of these creatures battle to cross the ACC as they migrate between the subantarctic islands and more temperate waters. Their strategy is to come to Antarctica to feast on the annual banquet that the ice melt lays on, then take off to warmer climes when winter sets in and the food supply dries up. Some of the seals and whales stay around the edge of the sea ice, which generally grows away from Antarctica during winter. The emperor penguin is the only large animal to stay behind and brave the Antarctic winter on land. More on these plucky birds in the box below.

Happy Feet or Cold Feet?

The emperor penguin is the tallest and heaviest species of penguin and is the only large animal to breed on land in Antarctica during winter (see Image #13). It breeds under the most extreme conditions of any vertebrate animal and as a result has developed some incredible biological and social adaptations to help it survive. We like to think penguins are like us, with their monogamy and romantic-looking courtship rituals, but in reality, if human men helped as much with child-raising duties as emperor penguins, there would be a lot fewer divorces.

In order that the chick will have grown enough to leave the nest during summer, emperor penguin pairs have to lay their eggs in the preceding autumn. The mothers are nutritionally depleted by egg-laying, so they are forced to head off to sea to feed. The female laboriously transfers the egg from the pouch beneath her stomach to the male, which nudges it onto the broad tops of his leathery feet and tucks his roll of belly fat over it to keep it warm. A dropped egg in the frozen conditions spells disaster. Then the males have to cope with looking after the egg without anything at all to eat for the duration of the furious Antarctic winter.

After the males have done the hard work of incubating the egg, the females return just in time to take the glory when the chicks hatch during the last of the winter darkness. Then spring rolls around and the parents take turns to head out to sea to find food for the kid. A feeding trip can include up to 1,200 dives over eight days, each up to depths of 400 metres. Only one in ten of these dives usually yields food. By the height of summer, the new penguin is ready to fledge and leave the nest, and the parents are pretty worn out.

During their time 'on the egg' as it were, the males fast for up to 115 days, losing up to 40% of their body weight in the process. During this time they have to bear temperatures of up to 50 below zero and winds of up to 200km/hr. To survive this, the emperor penguin has evolved plenty of fat stores and a lower body temperature – which is what made Happy Feet turning up on the comparatively tropical Kapiti Coast so unusual.

But the most intriguing part of the emperor penguin story is the social adaptation known as 'huddling'. Up to 5,000 birds can huddle together, with up to ten birds every square metre. Like a cycle peloton where individuals swap the lead to share the burden of overcoming wind resistance, the penguins in the huddle take turns on the outside to bear the brunt of the icy winds, before they are replaced and get to rest in the warmth of the centre. All this is done with an egg delicately balanced on their feet. Neat, huh?

In winter, the storms can't penetrate the sea ice to whip up the ocean's surface layer into a frenzy, so the water stays much the same temperature all year around. This helps avoid the boom and bust of food, and has given rise to one of the most diverse and rich benthic (seafloor) communities in the ocean. In fact, the seafloor around the Antarctic coast is the second most biodiverse ecosystem in the entire world, just behind coral reefs. And these aren't the sort of slow-growing organisms that you would normally expect in cold climates. They move quickly, as has been noted by the speed at which areas are recolonised after they have been scraped clean by passing glaciers. Incidentally, the periodic scraping it gets from glaciers is one reason why this environment is so dynamic and has such high diversity.

So what accounts for this incredibly rich life? Most of the life-forms on the seafloor are filter feeders, living off the goodness of the dead algae that rain down from above during the summer. So when there is a plankton bloom, anything that doesn't get eaten up by the animals grazing above drifts down and sustains the creatures below. Every time the tide goes in and out, this algae gets pulled off the seabed and floats around for filter feeders to pick them up. The algae stay fresh a long time, too, because the water keeps them cool all year round, a bit like the vege crisper in your fridge. Finally, instead of feeding on zooplankton as most other filter feeders do, the ones in Antarctica are uniquely adapted to eat phytoplankton. As we saw in the first section, this means that the food from the plankton bloom goes a lot further, because there are fewer links in the food web. As we saw, digestion is a wasteful process, with only 10% of the energy content of what's eaten adding to the waistline of the predator , so the fewer links in the food web, the more efficient the ecosystem.

Of course, once you get on land in Antarctica the sheer cold and isolation of the continent makes it one of the least biodiverse places on the globe. In fact, for diversity in land and freshwater systems, Antarctica is right at the bottom of the pile. All that can survive is a few fungi, bacteria and algae. Sure, the microbes are incredibly abundant and harbour some unique features – particularly in their ability to survive in the cold. But at the risk of raising the ire of Antarctic microbiologists, we're not going to bother talking about them.

As we saw early on, the lynchpin of the wider Antarctic ecosystem is krill. It is the major predator of tiny phytoplankton and the major prey of higher predators such as penguins, other birds, whales, fish and seals. But krill is a deep open-ocean species; on the shelf area of the Ross Sea, the role of krill is actually filled by the tiny Antarctic silverfish. Incredibly as it seems, the silverfish is a sprat-sized relative of the giant toothfish. It too has antifreeze proteins that allow it to survive life under the ice. The Ross Sea is subtly different from the rest of Antarctica in other ways, too, which we explore further in the box below.

The Ross Sea Ecosystem

The Ross Sea is covered by sea ice for most of the year. The ice expands from late February onwards, and retreats from late October. The algae growing under sea ice is important for silverfish, as it provides a food source for the animal plankton the silverfish eats over winter. The algae then blooms in the spring and summer along the coast, in polynyas (areas which do not freeze), and in the waters left open as the sea ice melts. There are Antarctic krill to the north of the Ross Sea, and smaller crystal krill in waters over the continental shelf to the south, but it is 15-25 cm long silverfish that comprises the bulk of the diet of almost all large predators in the area.

Silverfish pretty much get eaten by all animals in the Ross Sea, including fish, squid, seabirds and seals. There are large numbers of breeding emperor penguins and Adélie penguins in the Ross Sea. Several other species of birds breed in the region, including Antarctic petrels, snow petrels, and the south polar skua. Many other birds visit in summer, including two species of albatross. Seals are the most common marine mammals in the Ross Sea, including crabeater seals, Weddell seals, leopard seals and Ross seals. Baleen whales in the region include dwarf minke whales, Antarctic minke whales and smaller numbers of fin, humpback, sei and blue whales. Toothed whales sighted in the Ross Sea include killer whales, sperm whales, southern bottlenose whales and Arnoux's beaked whales.

As discussed, the seafloor of the Ross Sea is the most biodiverse part of the region (see Image #14). Falling detritus, and the growth of algae in coral (similar to tropical reefs), provide nutrition for the benthic (bottom-dwelling) creatures. These organisms include grazers such as urchins, sea cucumbers, and snails; predators, such as the Antarctic whelk and seastars; filter-feeders, such as Antarctic scallops, bivalves, anemones, soft corals, and sponges; and scavengers, such as large worms.

So that is a brief overview of the ecosystems in *Our Far South*. We will now focus on the species that are most at threat, and look at what those threats are. There are no surprises when you learn that these threats include many of the factors we have already examined so far – including the human race for resources and human-caused climate change.

What Species are Most at Risk?

This section is based on the *International Union for Conservation of Nature* (IUCN) red list of endangered species. If we include all threatened or 'near-threatened' species, there are thirty creatures on the red list that call *Our Far South* home. But before we take a look at those unlucky enough to make the top thirty, lets spare a thought for those that have no chance of making the list, endangered or not. There are simply too many species for the world's meagre science budgets to monitor.

For starters, some of the creatures are just too darn small or hard to find. Very few invertebrates (such as worms, spiders or insects) make it on the IUCN list, for example. It is also more difficult and expensive to study life in *Our Far South* because of the harsh conditions: after all, if you were a scientist, wouldn't you rather don khaki shorts, a pith helmet and chase butterflies in the tropics? We also do less science in the ocean, simply because this kind of research is so expensive. As a result, only 11% of ecological studies refer to the marine environment.[79] We haven't even seen most of the creatures that live in the mud on the bottom of the ocean, let alone counted them. Research funding is also swayed by public opinion. Cute, cuddly animals get researched far more than ugly, slimy ones. Some scientists think that up to 97% of the species on this planet could still be undiscovered.[80]

Even some pretty big animals slip beneath the radar because we don't know enough about them. Consider the beaked whales, for instance. These creatures have to come to the surface every 40 minutes or so to breathe, yet we know of their existence in New Zealand waters largely because they wash up dead on our beaches. We have no idea of the population size of any of the species of beaked whale that live in *Our Far South*, let alone how many species actually live there. So how are we supposed to know if they are endangered or if climate change or fishing may affect them? Incredibly, even relatively numerous whales such as Minke or killer whales have no reliable worldwide population estimate; we don't even know where they go to breed or if they are actually made up of distinct species – all of this despite all that 'research' effort the Japanese have been putting in. Whales are among the largest animals on the planet, and yet two new species have been described in the past decade.

The only way to redress this balance is by spending more on basic science, which is something that governments around the world are doing less and less. The pressure is on scientists to find ideas that have commercial applications. After all, why discover new *phyla* when you can invent a new nailfiler?

Rant over. Now to our not-so-illustrious top thirty. The red list includes ten albatrosses, five penguins, four whales, four land birds, three shags, two petrels, one shearwater, one sea lion and a partridge in a… wait. No, forget the partridge. See the table on the next page. For species that only exist in New Zealand (shaded in grey) we have used the Department of Conservation threat classification system. For the others, we have used the IUCN red list. They are listed in order of the severity of the threat status.

Threatened Species of Our Far South

Species	Antipodes	Auckland	Bounty	Campbell	Macquarie	Snares	Endemism (% World Popn in NZ)	Threat Status	NZ Population	World Population	Population Trend
Campbell Island Teal				X			100%	Nationally Critical	200		Increasing
Bounty Islands Shag			X				100%	Nationally Critical	618		Stable
New Zealand Sealion		X		X			100%	Nationally Critical	9,000		Decreasing
Yellow-eyed Penguin		X		X			100%	Nationally Vulnerable	4,136		Decreasing
Auckland Island Teal		X					100%	Nationally Vulnerable	600-2,000		Stable
Auckland Islands Shag		X					100%	Nationally Vulnerable	1,000-2,500		Stable
Salvin's albatross			X			X	100%	Nationally Vulnerable	61,500		Unknown
Antipodean Albatross	X	X		X			100%	Nationally Vulnerable	44,500		Decreasing
White-capped albatross	X	X					100%	Declining	190,000		Decreasing
Erect-crested Penguin	X		X				100%	Naturally Uncommon	160,000		Decreasing
Auckland Island Rail		X					100%	Naturally Uncommon	2,000		Stable
Campbell albatross				X			100%	Naturally Uncommon	49,000		Increasing
Campbell Island Shag				X			100%	Naturally Uncommon	4,000		Unknown
Snares Crested Penguin						X	100%	Naturally Uncommon	60,000		Stable
Southern Royal Albatross		X		X			100%	Naturally Uncommon	29,000		Stable
Snipe	X	X		X		X	100%	Naturally Uncommon	29,100		Stable
Buller's albatross						X	100%	Naturally Uncommon	64,000		Stable
Black-browed albatross	X			X	X	X	0%	Endangered	290	1,220,000	Decreasing
Blue Whale							0%	Endangered		1700*	Increasing
Fin Whale							0%	Endangered		15,178*	Unknown
Sei Whale							0%	Endangered		10,000*	Unknown
Royal Penguin					X		0%	Vulnerable	1,700,000		Stable
Grey-headed albatross				X	X		8%	Vulnerable	20,000	250,000	Decreasing
White-chinned petrel	X	X		X			6%	Vulnerable	200,000	3,500,000	Decreasing
Rock hopper Penguin	X	X		X			5%	Vulnerable	103,000	2,000,000	Decreasing
Wandering Albatross					X		0%	Vulnerable	20	16,000	Decreasing
Sperm Whale							0%	Vulnerable		150,000	Unknown
Sooty shearwater						X	50%	Near Threatened	10,000,000	20,000,000	Decreasing
Light mantled sooty albatross	X	X		X			31%	Near Threatened	18,178	58,000	Decreasing
Grey Petrel	X			X	X		27%	Near Threatened	106,000	400,000	Decreasing
Gentoo Penguin				X			0%	Near Threatened		520,000	Decreasing

Source: DOC threat classification system and IUCN red list.
* indicates Southern Ocean population

What stands out is the number of seabirds on this list. In fact, if you tot up all the albatrosses, petrels, penguins, shags and the shearwater, 21 out of the 30 species are seabirds. Seabirds are actually in peril all around the world – around half the species of seabirds are known to be experiencing population declines and almost 40% are on the threatened species list or close to it (like our near-threatened list above). New Zealand is home to more of these threatened birds than anywhere else in the world. We'll come back to the causes of this later.[81]

Some of the most startling cases of decline have happened in our own subantarctic islands.

We know this because some people were stationed on Campbell Island during World War II to watch out for German invaders. The Nazis never showed, but the coastwatch party kept themselves busy counting animals. Thanks to their records, we know that rock hopper penguin numbers on Campbell Island are down by 94% since World War II, while grey-headed albatross numbers have fallen by around 87%.

According to their records, the numbers of southern elephant seals on Campbell Island are also down by 97%. New Zealand has never had a large elephant seal population; nowadays there are 10 or so on Campbell Island and 250 on Antipodes Island. As we have seen this fall happened elsewhere but seems to have halted, so it doesn't make the IUCN list. But the story of the elephant seal population decline is an interesting illustration of how we have no idea what the 'natural state' of *Our Far South* is. By the time we started counting, the damage was already done.

Southern Elephant Seal[82,83]

There are many theories about the post-WWII decline in sea elephants (see Image #15). Climate change is a key suspect, which we will explore more in the next section, with the example of rock hopper penguins. But the more interesting hypotheses for the decline are all knock-on effects from whaling.

After whaling devastated whale stocks, elephant seal populations may have grown, partly as elephant seals replaced whales in their niche in the ecosystem. This could mean that the numbers of elephant seals when they were first counted in the 1940s were unnaturally high — up to 50% above their 'original' numbers. The population may have ended up 'overshooting' the food supply, particularly as whale stocks began to recover. What's more, without all the whales around to eat, killer whales and sharks may have turned on elephant seals as a tasty alternative.

On the other hand, the modern decline in elephant seals may be directly related to climate change. Changes in currents may be taking prey further away from the places elephant seals breed. Animals either have to dive deeper or swim further to find their food... and for animals that have their pups ashore, this means that they can supply less food to their pups, which means pup survival declines. Pesky climate change!

Now that we have some idea what species are at risk in *Our Far South*, it is time to look at the threats they face. Some of these are closely related to issues we have already looked at; namely the race for resources and climate change.

What are the Threats?

We like to think of *Our Far South* as untouched, but as we have seen, some aspects have already been significantly altered by human hands. Claims that the Southern Ocean and even the Ross Sea are a pristine environment are sadly anything but true. It began with sealing in the late 18th century, then moved to the great whales and attempts at farming in the twentieth century. But with the exception of the stubborn Japanese whalers, those days are gone, which is why numbers of the endangered blue whale are slowly recovering. Of course, that is not to say that the impact of whaling and sealing is not still being felt today, as shown by the box below.

The Krill Conundrum

Prior to whaling, the blue whale population consumed almost one million tonnes of krill every day. There were originally somewhere between 250,000-350,000 blue whales in the Southern Ocean; now there are around 1,700. So logically, there should be a lot more krill swimming around than there used to be. There isn't. Why not?

A few species of seals and penguins have benefitted from the decline in whales and seals, but there is evidence that this can't explain the gap, so it remains one of the great scientific conundrums about Antarctica. We don't yet know the answer, but finding an answer may require us to change the way we think about the environment.

Could whaling have actually reduced the abundance of krill? It seems bizarre, but it is possible, for a number of reasons. Remember that the productivity of the ocean depends on getting nutrients back to the surface, which relies on the ocean mixing. Incredibly, some scientists suggest that the 2 million whales removed by whaling might have had as great an impact on ocean mixing as the ACC itself, simply from the water disturbed from their swimming around.[84] Also the Southern Ocean is lacking in iron, and whale faeces are very high in iron (and other nutrients). So by diving down, eating and then pooping out the iron in their faeces (which float!), whales could have greatly increased the productivity of the Southern Ocean.[85] Despite what the Japanese say, having more whales may not mean less fish at all.

Nowadays, our top thirty threatened species face very different threats to the one that whaling once posed. As we saw, two thirds of our top thirty are seabirds; a recent study confirmed that the two greatest threats they face are from being caught as fishing bycatch and from invasive pests introduced into crucial subantarctic breeding habitat. Meanwhile, the insidious influence of climate change has had some proven impact but remains the big unknown, a threat for the future.[86]

Fishing

The real issue here is the knock-on effects fishing has on other animals, either by directly harming them or through indirect effects, such as removal of their prey. First, we will take a brief look at the impact from the fishing on the fish stocks themselves. For a more detailed look at fisheries management, we recommend reading our book *Hook, Line and Blinkers*. As with whaling and sealing, the early history of fishing was littered with horrendous failures. Antarctic rock cod, mackerel icefish and the Patagonian toothfish were all successively hammered from the 1970s to the late 20th century, and many of these stocks still haven't recovered. Mostly this was the impact of illegal fishing, which is still an issue in some areas.

Nowadays, most of the fish stocks in the Southern Ocean are managed pretty well. The deep water stocks within New Zealand waters, such as hoki, squid and southern blue whiting, are well managed, because they are regarded as an asset by the fishing companies themselves. It is in the industry's interests to keep the stocks abundant to ensure that they have a long-term, sustainable resource, and because abundance makes them easier to catch and keeps costs of harvest down. Some Southern Ocean fisheries are certified as sustainable fisheries by the Marine Stewardship Council, including krill, hoki and the Antarctic toothfish fishery (by NZ and UK fishers operating in the Ross Sea). The southern blue whiting fishery is also currently going through the MSC certification process.

So the problem with fishing in the Southern Ocean generally isn't the impact on the fish stock *per se*, but the impact on the wider environment. Fishing can harm our top thirty threatened species in two ways; gear can directly harm creatures, and the environment can also be affected by the removal of fish.

First, birds, seals and whales can all get killed inadvertently by fishing gear such as hooks, lines and nets. Let's take a quick look at the impact of fishing gear on our top thirty. Many seabirds, including yellow-eyed penguins, and some whales (like fin whales) can get caught in fixed fishing nets (or 'set' nets). Sea lions have in the past got caught in the nets of trawlers. Petrels, shags and shearwaters are all diving birds, so they can get snagged diving for fish caught in trawl nets or on longlines. Albatrosses don't dive, but they are still attracted by anything yummy on the surface, so they can also get snagged by longline hooks or their wings can get damaged when they hit the steel ropes dragging up trawl nets.

Some 50 million hooks are baited in New Zealand waters every year, and 35 species of albatross and petrel have been recorded caught on longlines in New Zealand since 1996. The most commonly caught species are white-capped albatross, white-chinned petrel and the grey petrel. This doesn't tell us much, largely because these are also the most numerous birds in our waters. The question is, what is the risk that fishing is damaging their population? This is a lot harder to work out, because we have to know how many birds there are, how many we are killing and how quickly they breed.

The reasons for the decline of seabird numbers are many: introduced and natural predators, loss of habitat, plastic debris — either harming them through ingestion or entanglement — oil spills or pollution, climate change, and fishing. The data aren't great, but New Zealand fisheries are *probably* reducing the populations of some of the seabirds on our top 30 list. A recent study suggests that a number of seabird species are being killed by interactions with fishing at a greater rate than they can breed, including the Salvin's albatross (which declined by 14% between 1997-2004) and the Antipodean albatross (more on these later). That really is an alb-atrocity.[87] Fishing also probably kills yellow-eyed penguins faster than they can breed, including from set-nets around the South Island. Given the impact of set nets on blue penguins, shags, Hector's and Maui's dolphins perhaps it is time to look at banning this fishing practice entirely.

A number of other species might be at risk from fishing – we just don't know enough. Black-browed albatrosses are declining around the world. These birds are certainly killed by interactions with New Zealand fisheries, but because very few of them breed in New Zealand, it is difficult to know to what extent New Zealand fisheries are responsible. Grey-headed albatross and light-mantled sooty albatross populations are declining around the world, but the data on interactions with New Zealand fisheries is too poor for us to know if we're really part of the problem. The impacts of fishing could also be high for Campbell and Buller's albatross, but these populations are not declining.

The populations of other seabird species are also declining, and fishing interactions could be partly to blame; including the white-capped albatross, white-chinned petrel, sooty shearwater and grey petrel. Were it not for fishing interactions, the population of the majestic southern royal albatross might be recovering. What we don't know is the number of our seabirds that get caught in other nation's fisheries. Judging by what happens in our waters (where fishing practices are probably better), it is probably a lot.

The second impact of fishing is the knock-on effect of removing fish from the ecosystem. Fishing a fish stock down to the so-called 'sustainable' level used by fisheries managers (usually 20-35% of the original population) means they can sustainably take more fish (the remaining fish grow faster and survive better), but it can still reduce the biodiversity of the environment – experience shows usually by around a third. Such unexpected consequences can happen because of predator-prey interactions between different creatures: fishing one animal means that the creatures that animal usually eats can proliferate, potentially creating an imbalance and causing mischief elsewhere in the food web. Another great example is that seabirds often depend on other creatures such as whales and tuna to round up bait fish for them – could our race for resources of whales and tuna have impacted seabirds?

So a well-managed fishing industry can still damage biodiversity, even if it doesn't make the fish itself extinct. The way to avoid this is by taking a holistic approach to managing the ecosystem. This is one reason why toothfish stocks are being managed at the precautionary level of 50% of the original population (rather than the more generally applied lower limit used in most other fisheries). It is also a strong argument for marine reserves – setting aside some parts of the ocean untouched, in case our meddling in the food web causes some unexpected side effects.

All these wider environmental impacts of fishing are best illustrated by the example of New Zealand sea lions. For many years, these threatened creatures have been declining, and it is difficult to know why. Is fishing to blame? And if so, is it the direct impacts of fishing or the fact that the sea lion is competing with the squid fishery for food?

Squid and sea-lions[88]

New Zealand sea lions are only found in this country and once ranged the entire New Zealand coastline. Most of their breeding now takes place in the Auckland Islands, where the females are known to wander miles inland to protect their pups; you can even run into them as you're strolling through the rata forest on Enderby Island (see cover image).

The population of sea lions in the main breeding colony (Auckland Islands) is declining. We know this because every summer since 1975, government officials (and volunteers) have wrapped up warm and travelled to the Auckland Islands to count them. Currently it is the job of the Department of Conservation. The number of pups born each year has fallen by 50% since 1998, and now probably only 9,000 animals remain in total (sea lions, that is, not DOC officials, although under the current government's cost-cutting regime, they seem to be endangered, too) which classifies them as a 'nationally critical' species. If current trends continue, the projection is for the sea lion to be extinct by 2035. The possible causes are many, including; fishing, disease, climate change, population overshoot and more.

Pup Production in Main Auckland Island Breeding Colonies

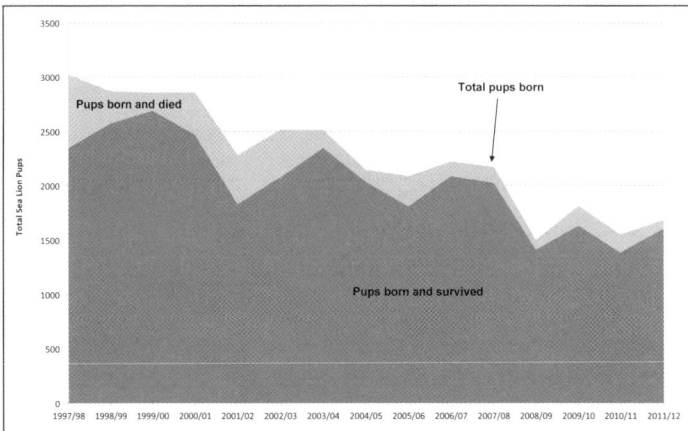

Source: Department of Conservation

Climate change could potentially affect food stocks, although this doesn't seem to be an issue for the smaller Campbell Island population, which is rising. Could climate change be impacting on one island and not the other? It's possible, but hard to be sure. Disease (campylobacter) struck in 1998 and wiped out 60% of pups, but this doesn't explain why the numbers of mothers has also been falling. One modelling study suggests that only fishing could explain the population fall.[89]

An important part of the sea lion diet is squid, and while a marine reserve was declared around the Auckland islands in the 1980s, it only extends 12 nautical miles, whereas the sea lions swim much further out to feed. Their main feeding grounds just so happen to be the best spots for squid fishing. In the past, some sea lions unfortunately got caught in trawling nets used by squid fishers. Environmental groups rightly complained, and eventually the Government and industry acted to manage the direct impact of the fishery on sea lions. The Government initially set a quota for acceptable, industry-wide mortality of sea lions: when this magic number was reached, everyone was obliged to stop fishing and come home.

The fishing industry argues that the new 'sea lion exclusion devices' (SLEDs) they use allow sea lions to escape their nets, so they are now having no impact on the sea lion population. Last year, there wasn't a single recorded death of a sea lion, and the fishing industry's tests (using crash test dummies!) indicate that SLEDs themselves don't harm sea lions. They argue something other than fishing must be to blame for the decline. Clearly the Ministry of Primary Industries agrees, as it is proposing to allow a higher catch of squid in the area.

Environmental groups, on the other hand, argue that this conclusion is premature. They question whether the SLEDs really work, as this fishing takes place in deep water where conditions differ from those in which the fishing industry testing was carried out. It could be that sea lions survive entrapment in the net and escape via a SLED, only to die outside the net and unobserved by researchers or fishers. After all, the extra minute or two sea lions spend finding their way out of the net in dark, deep water could run them out of breath. There is already evidence that they are operating at the edge of their natural range and at the limits of their physiological endurance: New Zealand sea lions are the deepest divers of all eared seals in the world.

But maybe sea lions simply can't get as much food as they used to, and that's reducing their numbers and their condition. It might be because of the impact of climate change on food abundance and location, or because sea lions are competing with the fishing industry for the same food: squid. There is certainly evidence of malnutrition among Auckland Island sea lions, and that the animals have to travel further for food. Is this because of the squid fishery?

Are we happy with sea lions sliding towards extinction? Should we aim to keep them at current levels, or should we aim to increase numbers? The Government has no population management plan for sea lions, which seems to be a real oversight. Once we draw this line in the sand, it becomes a matter of working out how to achieve the target. This might include experimenting with fishing practices to see if it could help. One approach that has been suggested is giving fishers the incentive to fish for squid outside the area where sea lions are known to forage.[90]

Neither the case of environmentalists nor that of the fishing industry seems conclusive as yet. As this book goes to print scientists from the Ministry for Primary Industries, Department of Conservation, fishing industry and WWF have agreed to work together to review the factors causing the decline of sea lions in the Auckland Islands. NIWA are also commencing a long-term study to explore the issue further, but in the meantime this raises the question of 'what should we do when we don't know the answer?'. At the moment, the fishery is the only major lever we have to slow or stop the sea lion from sliding to extinction. Given the evidence is not 100% clear cut, the precautionary approach would suggest we need to continue to manage fishing until we are certain it has no impacts. Based on the recent recommendations from the Ministry, this cautious approach simply isn't built into our fisheries management system. It remains to be seen whether the Minister shares this cavalier attitude, but it will be apparent when the squid fishing limit is finally set.

It's not all bad news. While many creatures are harmed by fishing, some do benefit. Some of the Buller's albatross populations have prospered as they have learned to scavenge the fish discarded by deep sea trawlers. The number of these birds at the Snares Islands almost doubled between 1969 and 1997, but have since plateaued. An analysis of their diet showed that they got around 60% of their diet from fish discarded from deep sea fishing vessels. We know this because stomach dissections revealed they were eating species albatrosses would normally not have access to, such as hoki and jack mackerel.[91]

Overall, however, it has to be said that there is strong evidence that fishing is causing some serious problems for many of our top thirty endangered species. We need to ask ourselves what we value more: a few million bucks and a few dozen jobs, or a bunch of animals that it is in our power to save.

Introduced Pests

So far, the part of *Our Far South* most affected by invasive species is the subantarctics, and most of those species were brought there by— you guessed it — humans. Pigs, cattle, sheep rabbits, and goats were all introduced by aborted attempts at farming or a desire to provide food for shipwrecked sailors. The other usual suspects, cats, rats, mice and rabbits have all found their way as stowaways or as part of thwarted human attempts at settlement. The introduced animals managed to survive in the wild, and continued to wreak havoc on the native wildlife and vegetation long after humans left. Even alien plants cause problems: they can spread quickly and out-compete local plants, reducing the number of species by as much as 40%.[92] Only the Snares and Bounty islands escaped the introduction of animal pests.

Humans introduced these species in the days when we didn't understand and appreciate the treasure trove of wildlife that the subantarctic islands carry. Now we realise the unique role these islands play as breeding outposts for many animals. As a result, we are now doing our best to eradicate pests from these islands, and to make sure no more invasive species can get here. All visitors to the islands must be accompanied by DOC staff, their footwear and clothes must be cleaned and bags and gear inspected, and no rubbish or food is allowed to be left. All ships making the trip have to be de-ratted and be free of any hull-encrusting organisms (such as undaria, a large, brown, invasive and aggressive Asian seaweed that has spread throughout New Zealand and that in 2006 invaded the Snares by hitching a ride on a fishing boat mooring rope).

So what pests are threatening our top thirty species, and where are they? The Snares and Bounty Islands have always been completely pest-free, as has Adams Island, the southernmost of the Auckland Islands group. Sheep, cattle, rabbits and mice were eradicated from another island in the Aucklands, Enderby Island, from 1993, and it too is now pest-free. Similarly, Campbell Island had cattle, sheep, cats and rats removed between 1984-2001: we'll discuss this shortly. Macquarie Island is currently the subject of another attempt at pest eradication: the Aussies are spending A$25 million over five years to eradicate rats, mice and rabbits there, and at last report, this effort seemed to be going according to plan. Goats were eradicated from Auckland Island in 1991 but cats, mice and pigs remain. Mice also remain on the Antipodes Islands. Interestingly, genetic testing of the mice on the Antipodes suggests they are more closely related to Spanish mice than New Zealand or British mice. They probably swam to the island following the wreck of a Spanish ship nearby.[93]

These animals — even humble mice — do untold damage to the top thirty threatened species of *Our Far South*, particularly seabirds like albatrosses and petrels, because they nest on or in the ground. Obviously pigs and cats are a huge threat to any species trying to nest on the main Auckland Island. We know for certain that these animals prey upon any young fledgling they find left unprotected. Of our top thirty, the following creatures nest on the main Auckland Island: the endangered yellow-eyed penguin, the rock hopper penguin, the southern royal, Antipodean and light mantled sooty albatrosses, snipe as well as the endemic Auckland Island rail, teal and shag. And who knows what else would nest on this, at almost 51,000 hectares, our biggest subantarctic island, if it were cleared of pests? The white-chinned petrel is everywhere on Disappointment Island in the Auckland Islands: they would almost certainly recolonise the main island if the pests were removed.

Rats eat the eggs and chicks of seabirds that nest on or under the ground, but rabbits and mice don't usually do such direct damage to wildlife. The major problem they cause comes from the changes they make to the habitat. Evidence from Antipodes Island suggests that mice damage the habitat by eating seeds and devastating invertebrate populations, such as that of the local weta (see cover image).[94] This can totally change the landscape, which makes it harder for many of our top thirty species to find good spots where they can make their nests and breed.

On Gough Island in the South Atlantic, house mice seemed to have evolved a particularly nasty streak which has seen them gang up to attack and kill the chicks of the endangered Tristan's albatrosses. Since this was first recorded, it is being noticed on other islands, and so far it seems more likely to happen where mice are the only introduced species on the island – as is the case in the Antipodes.[95] No such behaviour has been shown to happen on Antipodes Island as yet, but with up to 150 mice per hectare, they could well cause some disruption for some of the species that breed there, including: Antipodean, black-browed, white-capped and light mantled sooty albatrosses, erect-crested and eastern rock hopper penguins, grey (see the case study below) and white-chinned petrels, not to mention the snipe.

Grey Petrel[96]

Like most petrels, the grey petrel can dive deeper into the water for prey than the albatross — up to 10 metres deep. Petrels even have a nifty adaptation to take care of all the water that they end up swallowing as a result: a special part of their nose can excrete any excess salt. This diving behaviour is what gets them into trouble with fishers – even when the lines are weighed down, sometimes the birds can dive for the bait. As a result, the grey petrel is one of the most frequently caught birds in the New Zealand fishery. It is thought that some 45,000 birds may have been killed by the New Zealand tuna longline fishery over the past 20 years.

The largest breeding colony for grey petrels in the world is on our own Antipodes Island. Over 100,000 birds nest there every year, although numbers are difficult to calculate because they breed in burrows underground. This makes small, land-based predators like mice more of a problem. Not only can they harass the petrels directly (which hasn't been recorded on Antipodes as yet), but they also destroy the tussock habitat that the petrels love to burrow amongst. For this reason, the grey petrel would probably be the main beneficiary of the removal of mice from Antipodes Island.

See *www.milliondollarmouse.org* for plans to eradicate mice from the Antipodes.

At the moment, there are few established invasive species on Antarctica. Some non-native species have been found on the fringes, but it is too soon to tell if they will take hold. At the moment, the cold and the swirling current of the ACC seem to be barriers enough. Scientists predict that as the climate warms, more alien species will be able to colonise Antarctica by swimming or flying there, or travelling on ships, aircraft or rafts of kelp or plastic. Once again, tourists are a major risk.

Climate Change

The third major impact on our top thirty animals is our old mate, climate change. As we saw in the last section, this is causing changes to the winds and currents in the Southern Ocean, and (while it's not strictly a climate effect), increased atmospheric carbon dioxide levels are also making the water more acidic. Down in Antarctica, the ice is melting, and the prediction is that there will be less sea ice later this century; we're already seeing the retreat of the extent of annual sea ice on the Antarctic Peninsula. What does all this mean for our top thirty?

We know that climate change is likely to have a huge impact on wildlife, but evidence of the impacts so far is very difficult to come across. Climate change affects the whole environment in so many different ways that it can be very difficult to know what is causing change. The best example where we do have relatively clear evidence is probably that of the rock hopper penguin, as discussed in the box below.

Rock Hopper Penguins

Rock hopper penguins are a type of crested penguin. In recent years, scientists have even discovered that there are distinct rock hopper species: the imaginatively named eastern, southern, western and northern (you almost wonder what the scientists will do if they discover another species). Rock hoppers stand at 40cm tall and live in subantarctic islands right around Antarctica. The ones on our islands are the eastern species. As with most penguins, they are built more for swimming than dwelling on land, but as their name suggests, these plucky fellas will actually leap over obstacles on land rather than waddling clumsily around them. Like other crested penguins, they have those distinctive, spiky yellow and black feathers on their head, which look a bit like big eyebrows à la Groucho Marx.

In 1942, rock hopper penguin numbers on Campbell Island numbered some 1.6 million breeding pairs, but by 1985 they had dwindled to 103,000 pairs. There is no evidence that fishing or increased predation is to blame for the decrease. Scientists surmised that climate change had led to alterations in the currents around the island, and was affecting their food supply. Scientists from New Zealand's own National Institute of Water and Atmospheric research (NIWA) decided to test this theory.[97]

They have compared the carbon and nitrogen isotopes in feathers of modern birds and those in museums to see if the rock hopper's diet has changed over time. The nitrogen isotope gives clues as to where in the food web the rock hopper is feeding, and the carbon isotope tells us about the productivity patterns in the ocean.

Thanks to hunting, shooting penguin skin collectors of the 19th century and to museum collections around the world, the study was able to examine isotopes back to the 1880s. They found that the diet of the rock hopper hasn't changed – then as now, they are dependent on krill. What has changed is that since the 1940s, the ocean around Campbell Island has simply provided less food for the hungry rock hoppers. This is probably due to changes in water currents as a result of climate change. Warmer waters may be driving the prey of rock hoppers south or into deeper waters, making them harder to get, particularly during the breeding season. Understanding exactly what has driven the drop in productivity is currently the subject of further research.

Temperatures are unlikely to increase quickly enough to directly threaten any species in *Our Far South* in our lifetimes. Most species can handle a temperature change of 5-10 degrees without keeling over, and that rise shouldn't happen this century, even allowing for greater warming in the polar regions. Some animals can also move territory, although of course, this isn't always possible: many of our top thirty are tied to breeding in the subantarctic islands because it is the only land around. The greater threat is that warmer temperatures might allow new competitors and even predators to live in places they haven't lived before. This could completely change the environment in any area that warms. The best example of this is probably the potential for the loss of sea ice in Antarctica.

As we have seen, Antarctica's sea ice is a vital part of the environment. Sea ice provides shelter during winter, protecting the unique, complex environment that exists on the sea floor in the Ross Sea. This shelter also allows sea ice algae to grow, which nourishes the krill and silverfish during the lean winter months. Finally, when the spring melt comes, sea ice provides the nutrients for a bloom of algae and then the fragments of remaining ice provide a crucial habitat for the predators of krill and silverfish — penguins and seals — to use as a diving platform to feed off.

The Antarctic Peninsula is losing sea ice, and this gives us a unique insight into the changes that might confront the rest of Antarctica later this century, when the sea ice is predicted to reduce in size by a third. The signs aren't good: as temperatures have risen, Adélie penguins are being replaced by Gentoo and chinstrap penguins, the numbers of krill and silverfish are falling, and giant crabs are slowly moving closer to the delicate seafloor communities. We'll briefly look at each of these.

There are around 5 million Adélies in the world, with 40% of the population breeding in the Ross Sea area. These friendly little penguins are famous for bunching on the edge of the water, pushing one of their number in, checking to see if they fall into the waiting jaws of a killer whale or seal, before the rest all pile in behind. As the sea ice has melted on the Western side of the Antarctic Peninsula, numbers of Adélie penguins in the area have fallen by over 80% since 1975. They are used to feeding off the sea ice, but on the Antarctic Peninsula, they are being replaced by subantarctic Gentoo and chinstrap penguins which are used to operating in warmer, open waters and have lately been able to head south.[98] Could this be a sign of the future – true Antarctic seals and penguins getting pushed out by their subantarctic cousins? If so, where will the Emperor penguins like Happy Feet go?

Just as the Antarctic ecosystem depends on krill — or in the case of the Ross Sea, silverfish — krill and silverfish depend on sea ice. The algae that grow on the underside of the ice during winter provides krill and silverfish (particularly the babies) with much-needed food during the winter months. The sea ice also helps krill conserve energy, because instead of having to swim, they can crawl around on the underside of the ice. As sea ice has dwindled, this seems to have caused a decline in krill around the Antarctic Peninsula of between 38-81% (it is hard to get an accurate biomass on these).[99] Krill have been replaced in the ecosystem by jelly-like creatures called salps. The downside of salps is they are much lower in energy than krill, so this makes life much harder for whales that want to feed up before they swim back to the tropics. It's a bit like trying to carbo load for a marathon when all you can get your hands on is celery.

As for the seafloor, the most unique communities tend to live in shallow waters where the sea ice or ice shelf keeps the water a stable – 2°C year-round, such as under sea ice or the Ross Ice Shelf. As waters have warmed near the Antarctic Peninsula, giant crabs (spider and king crabs) have been moving closer. These are voracious predators that could potentially wipe out the seafloor communities if the warmer water allows them to survive there. We also expect that as the ice melts, more icebergs will be released, which can wipe out many benthic communities by scouring the seafloor. Finally as ice shelves retreat and collapse, these ecosystems will be completely lost, becoming more like the rest of the Antarctic shelf. This is clearly the greatest threat facing the unique environment on the bottom of the Ross Sea.

Another threat created by our carbon emissions is acidification. As we have seen, the cold Southern Ocean absorbs more than its share of CO_2, and this is turning the water acidic more quickly than anywhere else. By 2100, the levels of carbonate in the Southern Ocean are predicted to fall below saturation point, which means that there will not be enough in the water for organisms to use to build or repair their shells. It is too early to tell what will happen, but it is likely to make life harder for any life with a shell, such as the beautiful and tiny pteropods and the seafloor communities. Potentially acidification is a bigger, more immediate threat to the oceans than climate change, but as yet we don't know enough about it to be sure.

The final concern is what will happen to the productivity of the world's oceans with climate change. As we have seen, it's the winds and ice that create the currents that mix the ocean, recycling the nutrients to the surface so that algae can use them to grow. Through the ACC, the Southern Ocean plays an essential role here. In the future, there will probably be less sea ice and more wind in the Southern Ocean, so the effect on overall ocean productivity is ambiguous. There may well be the same amount of life in the global ocean as there is now.

Unfortunately, this isn't the case for the Southern Ocean itself, where it seems likely there will be less life overall. While the Southern Ocean may export the same quantum of nutrients to other seas in the world, algae in the Southern Ocean will be less able to use those nutrients themselves. The extra wind and reduction in sea ice will probably increase surface layer mixing in the Southern Ocean, meaning that algae will find it harder to stay on the surface where they have access to the sunlight they need to grow. Recent observations suggest that this is already happening, thanks to the stronger winds since the 1980s, the amount of algae growing in the Southern Ocean has fallen by a massive 10%.[100] That is 10% less food for all the creatures of *Our Far South* to eat. This will be a risk to all the creatures in our top thirty threatened species, particularly given everything else they face, as we can see with the Antipodean albatross in the case study on the next page.

Antipodean Albatross[101]

As with many environmental issues, the real problems are caused not by one of these impacts, but by the entire suite. The best example of a critter staring down the barrel of this effect is probably the Antipodean albatross (see Image #16).

The Antipodean wandering albatross is part of the great albatross family, with a wingspan close to three metres. In the past, the Gibson's albatross was considered to be a separate species, but recent evidence suggests it is sub-species of the Antipodean albatross. These creatures live for 40 years, and don't start breeding until they are ten years old, after which they will produce no more than one chick every two years. About two thirds of these chicks survive to leave the nest.

Antipodean albatrosses only breed in New Zealand's subantarctic islands. The largest colony is on the Antipodes Islands, but they also breed on the Adams and Disappointment Islands in the Auckland Islands group. There are about 45,000 Antipodean albatrosses left in the world, but their population has been falling. For example, numbers on Adams Island have declined from some 7,000 breeding pairs in 1973 to just over 3,000 there now. In the 1990s, scientists recorded around 3.2 nests per hectare, whereas 100 years ago an expedition reported finding 200 eggs from just five acres – that is roughly 100 nests per hectare. This fall doesn't seem to be slowing, either: over the last four years, Antipodean albatrosses have declined by 15% and Gibson's by 25%.

Many factors have contributed to this decline. A few intrepid pairs try to breed on the main Auckland Island, but most of their chicks are snapped up by the wild pigs and cats there. Thankfully, Campbell Island now no longer has this problem, so hopefully the bird will start recolonising Campbell Island in numbers. There is no evidence that the mice on Antipodes are hassling the eponymous albatross, but they may well be. Mice are also known to damage important breeding habitat.

Fishing is another big problem. In the past, tuna longliners killed large numbers of Antipodean albatrosses because they used squid as bait – which happens to be the birds' favourite food. Things have improved recently (as we will see later on), but as far as we can tell, New Zealand fishers are still killing enough adults to stop the population from rising.[102] Another huge concern is that these birds are global citizens. The albatrosses breeding on Antipodes tend to fly to South America to feed, and fishing practices are less enlightened over there. The Chilean swordfish fishery in particular probably kill large numbers: they don't keep good records, so it is hard to know. They're catching our birds, over there.

Meanwhile, the albatrosses that breed on Adams Island (the so-called Gibson's sub-species) face another problem. They fly across the Tasman to feed while they are breeding. Not only do they have to contend with the Australian and Kiwi fisheries as they do this, but there is also emerging evidence that they are struggling with the massive change in currents happening off the east coast of Australia (as we saw in the climate change chapter).

Climate change means that the distinctive life we find in *Our Far South* may look more and more like the life we find in northern climes. The beasties most at risk from this change are the ones that are sensitive to the temperature change but can't easily pack up and move home, or are having their home territory squeezed. Polar bears standing on shrinking icebergs are the poster child of climate change: all the creatures that call *Our Far South* home could just as easily be used as the face of doom instead.

Other Risks

Of course, introduced pests, climate change and the race for resources are only three of the major threats to the wildlife in *Our Far South*. This section will briefly cover other risks, such as the build-up of pollution in the atmosphere and plastic in the ocean. Habitat change is also a major issue, but this has already been covered in the race for resources section. We won't look at natural risks such disease (although humans can transmit some diseases to wildlife, and vice versa) and natural predators such as skua, sharks or killer whales.

Increasing levels of carbon dioxide in the atmosphere is not the only scourge of our making affecting *Our Far South*. We have already seen some of the examples of pollution from ships and research stations operating in Antarctica. What is harder to believe is that the atmosphere has carried pollutants from all around the world to Antarctica.

'Persistent Organic Pollutants' sounds like a door-to-door salesman flogging hippy food, but in reality, they are even more persistent and far more dangerous. Organic pollutants (organic as in containing carbon, not sandal-wearing) include DDT (once used in agriculture), PCBs (used as coolants by factories) and PBDEs (flame retardants). They are toxic and don't break down for a long time, so can build up in the food supply. This means that if you eat something with some organic pollutants in it, all those pollutants will stick around in your body, mostly building up in fat cells. The longer an animal lives, the more these chemicals can build up, and they can start to affect the animal's ability to breed. Seals and whales can even pass these chemicals to their offspring through mother's milk, which typically has a high fat content. Of course, this becomes a real problem for things at the top of the food chain, including seabirds such as the gentoo penguin and toothed whales like the sperm whale, which are both on our top thirty list.

Because of their nature, organic pollutants tend to evaporate in warm regions and become part of the atmosphere, before condensing once they've been blown to cold regions. Once they condense in a cold region, they don't tend to evaporate again, so they accumulate there. Once deposited, there is nothing that can get rid of them: we just have to wait until they eventually break down. Just like heavy fuel oil, which we discussed in the race for resources section, the breakdown of long-lived organics is likely to take even longer in the cold environment.

These worrying properties have led to organic pollutants being banned or progressively phased out around the world. Nevertheless, they have found their way in significant quantities to Antarctica. This is possibly partly due to some leaching from ships and research bases, but it is more likely due to the pathway described above: that is, they have evaporated and were transported around the world in the atmosphere. They ended up becoming part of the Antarctic snow, and are ending up in the coastal ocean, where they build up in the biomass.

In our book on fishing *Hook, Line and Blinkers,* we looked at the threat that plastic poses to the ocean. We saw how this ubiquitous feature of modern life is so damaging to the ocean for the very reasons it is useful on land: it is cheap, light and durable. It doesn't so much break down as break up into smaller and smaller bits, bits that animals can easily mistake for food. Tests have found a quarter of seabirds have plastic in their stomachs, and there is evidence that this is a real problem for some of our top thirty, notably royal penguins and wandering albatrosses.

Another source of plastic pollution actually occurs as a side-effect of fishing. When nylon nets get lost or damaged, they often drift through the ocean, continuing to fish until they fill up with fish and sink to the bottom. When the fish get eaten (or rot), the load is lightened and off they go again… Again, lost and discarded fishing materials such as nylon and rope debris are a major threat for whales, birds and seals.[103] As we saw, an estimated 200,000-300,000 cetaceans (whales and dolphins) are accidentally killed each year by nets. This is mostly the little guys, but nets can even pose a threat to all of the big whales in our top thirty, like fin whales.

Once in the ocean, it doesn't matter where the plastic came from: both populated areas and remote islands seem to be awash with rubbish. Plastic rubbish is buoyant, so it will float just about anywhere the current takes it. Most people have heard of the Pacific garbage patch, but plastic can also make its way to *Our Far South*. Thanks to the Antarctic Circumpolar Current, any plastic in the Southern Ocean quickly makes a trip around the world. This is why footballs from a wayward kick in South America can end up washing up on the beaches of subantarctic islands, as we explore in the box below.

Macquarie Island Rubbish Study[104]

While monitoring weather patterns on Macquarie Island, the Aussie scientists started collecting the rubbish that washes up in Sandell Bay (see Image #17). This rocky bay is on the west coast, slap bang in the path of the ACC. This wasn't your average beach clean-up: in the meticulous way of scientists, they sorted all the rubbish they found (right down to pieces 1cm small), categorised it and kept track of it over time. The results are fascinating.

Between 1991 and 2001 (the time that rubbish was being collected), the amount of rubbish collected didn't change that much, but its consituents did. The proportion of rubbish that was plastic grew from 80% to almost 99%. And over time, more and more of that plastic was fisheries-related, including nets, ropes, bait straps and floats. Some of it came all the way from the Falkland Islands, a trip that would probably take just over two years. The moral of the story is be careful what we put in the Southern Ocean, because the ACC could take it anywhere.

As with so many of the issues facing our ocean, we just don't know how much of an impact we are having with our callous use and casual disposal of pollutants and plastics. We think we understand when we see an albatross choking on a piece of plastic or a sea lion strangled by a fishing rope. But the really scary thing is that we don't even know about everything that is living in the ocean yet: how can we possibly understand the full impact?

What Can we Do?

So far, this section has been all doom and gloom. Now it's time for the good news. In this section, we will look at what can be done to conserve the species of *Our Far South*. Unfortunately, sorting climate change would take a whole book in itself (and that's the plan). Issues like the threat of mining south of 60°S are not really a 'live' issue: they simply require staying aware of what is happening in Antarctica and the subantarctic islands. The issues we will focus on here will be basic science, fisheries management, marine protection and pest eradication.

Basic Science

One of the consistent themes of this book is that we just don't know enough about our impact on the environment through climate change and the race for resources. As a result, we don't know anything about some creatures that might be endangered, and for the ones we do know about, the information is quite patchy. As we have seen, thanks to a quirk of international politics, the scientific effort in Antarctica is quite good. Indeed, it is so good that it shows up our own lack of effort in the subantarctics.

The ocean, climate and life of the subantarctic region directly affects us, influencing everything from fishing to our weather. We need to look more closely at issues such as changing ocean circulation, ocean acidification, changing wind patterns, changing fisheries and declining iconic species. Yet most Kiwis barely know that the region exists and worse still, we have no coordinated national research effort in the region. At the moment, the sheer cost of operating in the region is too high for a scientist to propose working alone.

Some Kiwi scientists are working with DOC to scope the idea of establishing a subantarctic research station, perhaps on the Auckland Islands. This could provide a key focus for scientific effort and growing the understanding as well as engaging Kiwis in that part of the world. If it goes ahead, this could help bring science to the fore, increase national understanding, value and commitment to a New Zealand World Heritage site, as well as an important part of the global ocean system on our back doorstep.

With all basic science, the challenge is finding someone to pay for it. The Government is very focused on 'innovative' science with a commercial application. Monitoring can be perceived as neither innovative nor commercially applicable, but that doesn't mean it isn't important. After all, it may help us protect the very planet that we live on. And without that, all our lucrative innovation will be of precious little use.

Fisheries Management

Improved fishing practices have in recent years done a lot to reduce the industry's environmental impact, particularly in New Zealand's Southern Ocean.

Southern Seabird Solutions & CCAMLR [105]

Southern Seabird Solutions is an alliance spanning government, fishing companies and environmental groups, working to promote better fishing practices and to help avoid seabird capture. The group has made huge gains, although there is more to do. Southern Seabird Solutions focuses on working with fishers, as these guys often have the experience and ideas to suggest practical ways to reduce bycatch, never mind that it is these guys who will need to actually put these ideas into effect.

New fishing techniques have been developed aimed at reducing seabird deaths associated with fishing, such as fishing at night, dyeing the bait blue, weighting lines so that they quickly sink beyond the limited dive range of petrels and albatrosses and other techniques such as streamers to scare the birds away. While this progress is encouraging, so far there is no evidence that seabird capture rates have fallen — in some fisheries the number of birds caught has dropped, but this may be simply due to the fact that there is less fishing than there used to be.

Perhaps the most successful example of managing seabird deaths is fishing around Antarctica, which is tightly regulated by CCAMLR. They require full observer coverage of the fleet and these closely monitor seabird deaths – boats that catch more than 3 birds in a season are sent home. The New Zealand toothfish fishery in the Ross Sea has had no bird deaths at all in fourteen years of operation, which is quite an achievement even considering that there are fewer birds that far south. The secret of this success seems to be the ban on discharging offal from the vessel in CCAMLR areas, which have prevented seabirds from being attracted to vessels. If all vessels learnt this lesson and only discharged offal when there are no dangerous nets or hooks around for birds to get caught in, this might help address the huge problem of seabird deaths.

While New Zealand is acting to reduce our fishing impact domestically, these birds go many places in the Southern Ocean to feed, so fishing practices used by South American nations — and indeed, any nation fishing in the open sea — are also important. Wandering albatrosses such as the Antipodean are most at threat. They routinely travel thousands of miles, foraging as far afield as the Chilean coast. So we need to work with other nations to stop them catching 'our' birds over there. Now Southern Seabird Solutions are working with South American fishers to improve their practices, but it is tough going because we can't force them to adopt best practice. All we can do is encourage and educate.

Marine Protection

In the Race for Resources section, we have already looked at the lack of marine reserves and the absence of precautionary standards surrounding ocean drilling (particularly in the big waves of the Southern Ocean) in our EEZ. These issues are currently being looked at by the Government, but as we noted, no one has overall responsibility if things go wrong.

Exactly how much area should be set aside for marine reserves? You will recall the argument for marine reserves as an insurance policy: given all the other impacts we have on the ocean, it is useful to set some parts aside in their pristine state in case we stuff the rest up. This makes the oceans as a whole more resilient in case ocean ecosystems start collapsing. So the extent of reserves we have reflects how much risk we are prepared to take, and in part, how intensively we exploit resources elsewhere.

The theory with MPAs is not to lock up the whole area but to protect the areas with particular ecological significance. It is a similar approach to what we do on land with our National Parks and natural reserves. So why the calls to lock up most of the Ross Sea region (remember, this extends from the Ross Sea out to 60°S)? The argument really is that the Ross Sea is a 'pristine' marine ecosystem, and in the context of marine protection in Antarctica, it is easier to protect this whole area than trying to rejuvenate an already depleted ecosystem elsewhere. Having a pristine area can also create a laboratory for scientific studies — it can be compared with disrupted ecosystems elsewhere. The claim that the Ross Sea is pristine comes from some analysis of human impact on marine ecosystems,[106] but this work has been criticised because it doesn't include the impact of whaling and sealing, which precipitated a massive change in the Antarctic environment. That would be like wiping out the elephants and hippos from Africa and nevertheless calling it pristine.

There is also an argument that because the Ross Sea is such 'a special place', it should not be violated. This sort of environmental platitude tugs at the heart strings but doesn't really help much in deciding how much marine protection is needed. All places on earth are special in some way or another: you might just as well argue that we should protect an area because it is 'pretty'.

The only way to be sure about how much area of marine reserves we need is to base decisions on science; an approach known as systematic conservation planning. This process gathers all the data on important ecosystem features and possible threats and seeks to protect portions of those ecosystem features with an emphasis on the most ecologically important areas. It does this while simultaneously balancing resource extraction activities like fishing. It also gives the marine protection a science-based reason for being, which in turn allows for scientific monitoring to assess whether the MPA is actually doing what it is supposed to. This is the exact approach that has been used to create the draft New Zealand proposal for marine protection in the Ross Sea.

As an example the New Zealand process tried hard to segregate the fishery from the predators that it might compete with (Weddell seals and Type C killer whales). The Ross Sea killer whale is a new species (only identified in the last 5 years) which specialises on hunting toothfish. They only hunt in particular places because they rely on ice edges and they can't dive very deep, whereas fishing boats can fish in deeper water. So it is possible to segregate the fishery from the killer whales; the predators operate in one area, the fishery operates in another.

As we saw in the Race for Resources section, the work to create a marine protected area in the Ross Sea region is well underway. The New Zealand draft proposal would, if adopted, protect the vast majority (95%) of the Ross Sea itself, which is the area that has the greatest ecological significance. Even outside the Ross Sea, the New Zealand proposal would protect

30-40% of the wider Ross Sea region, which is still excellent by international standards. If the New Zealand proposal were approved, most of the fishing would take place outside the Ross Sea where ecosystem effects are expected to be lowest. As we have seen, this fishing fleet could also assist in research and monitoring of the area, as well as deterring illegal fishers from operating in the area.

As we saw in our book on fishing *Hook, Line and Blinkers*, the real travesty is the lack of protection of the ocean within our own EEZ. By contrast, Antarctica is shaping up as a good news story. Compared with Australia, very little of our ocean is protected. Currently, we don't even have the legislation needed to create marine reserves outside our 12-mile zone. Australia does: that's why a third of the waters around Macquarie Island are protected right out to the edge of the EEZ. We need to follow Australia's lead and implement legislation, followed by a Government-led process of zoning the ocean (including setting up reserves). This could potentially make a big difference to the wildlife of *Our Far South*.

Another issue where our approach in Antarctica puts our domestic position to shame is on fish stocks. You will recall that the adult toothfish population is managed at the precautionary minimum level of 50% of their pre-fishing numbers. By contrast, many of our fish stocks, particularly the inshore ones that are being competed for by commercial and recreational fishers, are fished right down to a Maximum Sustainable Yield of 20 to 35% (that is, to the point where the fish stock generates the most new adults). This is a risky approach that leaves little room for error in the fisheries data, and just ignores the ocean ecosystem. A more precautionary approach is needed, which has been recognised in recent Ministry of Fisheries harvesting targets (30-40%), but it remains to be seen if these are actually implemented in practice.

Pest Eradication

On land, our subantarctic islands already have the highest possible level of protection. They are all nature reserves and access is monitored and restricted by the Department of Conservation. The idea of this is to protect the islands from adverse impacts – particularly anyone introducing alien species (even by mistake – imagine a spore on a shoe or pregnant rats on a ship) which could threaten the local wildlife. In December 1998, the islands were listed as a UN World Heritage Area. This recognises the role these islands play in protecting biodiversity and offering a refuge for life in the Southern Ocean, and puts them on a par with the Grand Canyon and Great Barrier Reef.

But this status does not undo the damage already done to their ecosystems. As pests arrived and habitat was destroyed, we lost species. On Campbell Island, parakeets were eliminated completely before they were even discovered: we only know they were there at all from skeletal remains. Teal, snipe and a number of small seabirds were eliminated from the main island and are only found on small offshore islands and rock stacks, making their survival precarious.

The predator-free islands are hugely valuable examples of nature as it was before humans came along. Over time, New Zealand has developed the know-how to remove some of the introduced pests and help restore these islands to approximations of their natural state. This provides more space for creatures like albatrosses to breed, which might allow their populations to recover. This will also allow the islands and their precious cargo of species to better handle the other threats that humans are sending their way, such as climate change and pressure from fishing. Perhaps one of the greatest examples of our skills in this area was the progressive eradication of pests from Campbell Island, which we explore in the case study on the next page. Renowned scientist Tim Flannery even rated this achievement above the successes of our dairy industry and the All Blacks!

Campbell Island Regeneration

Since the Campbell Islands were cleared of pests, the regeneration of the island has moved quickly. The removal of the sheep in 1991 allowed the megaherbs to start growing back. The lack of dry open spaces for stalking obviously made life too difficult for the cats, as they started dying out soon afterwards.

But the biggest challenge by far was the removal of Norway rats. The previous biggest island cleared of rats was Kapiti, at 2,200 ha, whereas Campbell Island was five times larger at 11,000 ha. In 2001, DOC began an ambitious rat poisoning programme involving 19 personnel and five helicopters. Some 120 tonnes of bait was spread by helicopters, when the Southern Ocean winds calmed enough for them to fly, at a total cost $2 million. The island had to be split into grids and helicopter pilots used GPS to make sure that bait ended up within the home range of every rat. Success was achieved when the island was declared rat-free in 2003, making it the largest island in world to be cleared of rats, ever.

So far, the recovery from sheep and rat eradication has been outstanding. The megaherbs are growing back and their flowering in mid-summer is spectacular. In fact, all of the vegetation is showing positive recovery. The insect life is now thriving again with species such as weta becoming obvious everywhere. Land birds such as the endemic pipit, the rare flightless teal and the snipe were all previously restricted to very small numbers on offshore islets, but have now re-colonised the island. The seabirds are also coming back – species such as diving petrels, storm-petrels, white-chinned petrels, and sooty shearwaters are now all breeding on the main Campbell Island.

It is hoped the programme will benefit 33 species, including many of our top thirty: the teal and snipe, Campbell albatross, grey-headed albatross, grey petrel, yellow-eyed penguin and rock hopper penguin. In fact, the Campbell albatross, teal and snipe are already showing signs of recovery, and the teal may soon move off the Nationally Critical list.

The snipe is a particularly interesting tale. In 1997, a group of scientists were looking for teal on the small Jacquemart Island off the south coast of Campbell Island. Their bird dog flushed a bird out of the bushes that they had never seen before. This turned out to be the Campbell Island snipe, the first new bird subspecies discovery in New Zealand for over a century.

In 40-knot winds, the scientists searched the island, finding 10 snipe in total. They are tiny birds, fitting into the palm of your hand. They live on the ground and can fly when they have to, but not very far. Now that Campbell Island is cleared of cats and rats, the snipe and teal are rebuilding their populations on the main Campbell Island.

New Zealand's Department of Conservation is seen as the world leader in rodent eradication operations from offshore islands, having carried out over 150 rat or mouse eradications on islands around New Zealand. DOC has also exported its expertise to countries such as the USA, Britain, and many South Pacific nations.

DOC has eradicated a number of pest species from the islands in the subantarctic and has recently assisted the Australian Government, in the form of the Tasmanian Parks and Wildlife Service, to carry out the eradication operation for rabbits, ship rats and mice on Macquarie Island, 12,785 ha in area and 1,500 km from Tasmania. The five-year, A$25m Macquarie operation showed that there can be downsides from pest eradication, as was seen when 2,000 birds died following the commencement of the eradication programme. It appears the birds ate the bodies of the animals that were poisoned and became ill themselves. This is unfortunate: but ultimately more birds will be saved from the eradication in the long run.

So assuming Macquarie's project is successful, the only remaining animal pest species on the subantarctic islands will be mice on the Antipodes Islands and pigs, cats and mice on the main Auckland Island. The methodologies to remove these pest species are already developed; the only barrier is money. To eradicate mice from 2,025 ha Antipodes Island (850 km from NZ) would cost about $1m with logistical support from the Navy. The official estimate is that to eradicate both pigs and cats from 50,990 ha Auckland Island, it would cost $20m. Mice on the Auckland Islands are also a long-term target, although that will be extremely difficult given that it is five times the size of Campbell Island.

As a lasting legacy from the *Our Far South* voyage, I (Gareth) decided to work with DOC and the New Zealand public to eradicate mice from the Antipodes. This Million Dollar Mouse campaign will leave the island completely pest-free, providing a haven for the precious wildlife that lives in the area. This would be a great achievement, although the challenge of eradicating pests from the main Auckland Island is truly the pest eradication holy grail.

Summary

Our Far South contains a treasure trove of precious species. Some of them are endangered, others we barely know anything about and yet many face threats, including from fishing and introduced pests. Longer term, climate change is also an issue. Plants are the basis of all life, and in the Southern Ocean algae are already struggling from a lack of iron and too much churn. Climate change could make that worse by increasing the wind and reducing the sea ice in the region. We need to do more to understand the region, but in the meantime there are things we can do. We need to protect more of our ocean, improve fishing practices, and clear introduced pests off our subantarctic islands so that the species living there can breed and feed uninterrupted.

Sorry greenies, but these are far greater conservation priorities than worrying unduly about the toothfish fishery in the Ross Sea.

Thanks for coming on this journey with us. If you would like to help make a difference to *Our Far South* you can contribute to the Million Dollar Mouse project at *www.milliondollarmouse.org*.

ENDNOTES

1   Hamish Campbell and Gerard Hutching (2007) *In Search of Ancient New Zealand*. Penguin (NZ)

2   Chilvers, L. Foraging Locations of Female New Zealand Sea Lions From a Declining Colony. *New Zealand Journal of Ecology* (2009) 33(2): 106-133

3   Carter, L., McCave, I.N., Williams, M., 2008. Circulation and Water Masses of the Southern Ocean: a review. in (Florindo, F., Siegert, M. editors) *Antarctic Climate Evolution, Developments in Earth and Environmental Sciences 8*. Elsevier, Amsterdam, 606 pp.

4   Snodgrass, F.E., Groves, G.W., Hasselmann, K.F., Miller, G.R., Munk, W.H., Powers, W.H., 1966. Propogation of ocean swell across the Pacific. *Philosophical Transactions of the Royal Society of London* 259, 431-497.

5   Rintoul, S et al (2012) *The Southern Ocean Observing System: Initial Science and Implementation Strategy*. SCAR

6   Rintoul, S et al (2012) *The Southern Ocean Observing System: Initial Science and Implementation Strategy*. SCAR

7   McNeil, B (2008) *Global Ecology of the Oceans and Coasts* in Patterson & Glavovic *Ecological Economics of the Oceans and Coasts* Edward Elgar, UK

8   Patterson, M(2008) *Towards an Ecological Economics of the Oceans and Coasts* in Patterson & Glavovic (eds) *Ecological Economics of the Oceans and Coasts* Edward Elgar, UK

9   http://earthobservatory.nasa.gov/Features/CarbonCycle/carbon_cycle4.php

10   http://www.coolantarctica.com/Antarctica%20fact%20file/antarctica%20fact%20file%20index.htm

11   Anderson, John B. (1999). *Antarctic marine geology*. Cambridge University Press. p. 59. ISBN 0521593174.

[12] Manighetti, B., 2001. Ocean Circulation: the planet's great heat engine. *NIWA Water and Atmosphere* 9, 12-14.

[13] Rhys Richards (2003): New market evidence on the depletion of southern fur seals: 1788–1833, *New Zealand Journal of Zoology*, 30:1, 1-9

[14] Bill Mansfield. 'Law of the sea', Te Ara – the Encyclopedia of New Zealand, updated 3-Dec-09 URL: http://www.TeAra.govt.nz/en/law-of-the-sea/3/1

[15] Peat, N (2003) *Subantarctic New Zealand: A Rare Heritage.* Department of Conservation, Invercargill.

[16] http://www.nzherald.co.nz/nz/news/article.cfm?c_id=1&objectid=10415310

[17] Dudeney, J.R. & Walton, D.W.H Leadership in politics and science within the Antarctic Treaty. *Polar Research* 2012, 31, 11075, DOI: 10.3402/polar.v31i0.11075

[18] Tin, T. et al. 2009. Impacts of local human activities on the Antarctic environment. A review. Antarctic Science 21(1): 3-33. doi:10.1017/S0954102009001722

[19] Quote from Prem Chand Pandey, former director of NCAOR (National Centre for Antarcti and Ocean Research, Indian Government. Taken from Jayaraman, K.S. India plans a 3rd Antarctic base. Nature, 447, 3 May 2007.

[20] Fogerty, E (2011) *Antarctica: Assessing and Protecting Australia's National Interests.* Lowy Institute policy brief, http://www.lowyinstitute.org/Publication.asp?pid=1661

[21] Fogerty, E (2011) *Antarctica: Assessing and Protecting Australia's National Interests.* Lowy Institute policy brief, http://www.lowyinstitute.org/Publication.asp?pid=1661

[22] Fogerty, E (2011) *Antarctica: Assessing and Protecting Australia's National Interests.* Lowy Institute policy brief, http://www.lowyinstitute.org/Publication.asp?pid=1661

[23] Russian Federation (2011) *On strategy for the development of the Russian Federation activities in the Antarctic for the period until 2020 and longer-term perspective.* Antarctic Treaty Consultative Meeting 34, Working Paper 55

[24] http://rt.com/news/arctic-lead-cold-war-667/

[25] http://www.nbr.co.nz/category/category/great-south-basin

[26] Aronson, R.B. *et al* Anthropogenic Impacts on Marine Ecosystems in Antarctica. Ann. N.Y. Acad. Sci. 1223 (2011) 82-107.

[27] Aronson, R.B. et al Anthropogenic impacts on marine ecosystems in Antarctica. *Ann. N.Y. Acad. Sci. 1223* (2011) 82-107

[28] Turner, J et al (2009) *Antarctic Climate Change and the Environment,* Scientific Committee on Antarctic Research, Cambridge, UK

[29] Morikawa, Jun (2009). *Whaling in Japan: Power, Politics, and Diplomacy.* Columbia University Press. pp. 29–30.

[30] Ivaschenko, Y.V. *et al Soviet Illegal Whaling: The Devil and the Details.* NOAA, USA

[31] Read, A.J et al. Bycatch of Marine Mammals in U.S. and Global Fisheries. In *Conservation Biology* (2006) Volume 20, No. 1, 163–169

[32] Ministry of Fisheries, (2008) *Harvest Strategy Standard*, Ministry of Fisheries, Wellington

[33] http://www.fish.govt.nz/en-nz/International/ Fishing+in+the+Ross+Sea.htm

[34] http://www.iwatchnews.org/2011/10/02/6745/spain-doles-out-millions-aid-despite-fishing-companys-record

[35] http://www.nzherald.co.nz/nz/news/article.cfm?c_id=1&objectid=10743959

[36] Doran, P.T. et al Examining the Scientific Consensus on Climate Change in *EOS* Vol 90, Number 3, p22-23 (2009)

[37] Edward J. Larson & Larry Witham **Leading scientists still reject God** *Nature* 394, 313-313, 1998

[38] Hansen, J., A. Lacis, R. Ruedy, and M. Sato (1992), Potential climate impact of Mount Pinatubo eruption, *Geophys. Res. Lett.,* 19(2), 215–218, doi:10.1029/91GL02788.

[39] Source: United States National Oceanic and Atmospheric Administration http://www.ncdc.noaa.gov/oa/climate/ globalwarming.html

[40] Source: NIWA

[41] http://ipcc.ch/publications_and_data/ar4/wg1/en/faq-6-2.html

[42] Millennium Ecosystem Assessment (2005) *Ecosystems and Human Well-being: Synthesis.* Island Press, Washington, DC.

[43] Shakun, J.D. et al Global warming preceded by increasing carbon dioxide concentrations during the last deglaciation in *Nature* Volume: 484, Pages: 49–54 Date published: (05 April 2012) DOI:doi:10.1038/nature10915

[44] Bernstein, A et al (2007) *Climate Change 2007: Synthesis Report.* Intergovernmental Panel on Climate Change

[45] Hoegh-Guldberg & Bruno (2010) *The Impact of Climate Change on the World's Marine Ecosystems.* Science Vol 328

[46] Bindoff, N.L. et al (2011) *Position Analysis: Climate Change and the Southern Ocean.* Antarctic Climate and Ecosystems Cooperative Research Centre.

[47] Bindoff, N.L. et al (2011) *Position Analysis: Climate Change and the Southern Ocean.* Antarctic Climate and Ecosystems Cooperative Research Centre.

[48] Nicholls, R.J. Sea-Level Rise and Its Impact on Coastal Zones. In *Science* Vol 328, 18 July 2010

[49] Turner et al (2009) Antarctic Climate Change and the Environment Scientific Committee on Antarctic Research, Cambridge, UK

[50] M. Susan Lozier, et al. Deconstructing the Conveyor Belt. *Science* 328, 1507 (2010). DOI: 10.1126/science.1189250

[51] Kerr, R.A. Ocean Acidification: Unprecedented, Unsettling. In *Science* vol 328, 18 June 2010

[52] Moy, A.D. et al (2009) *Reduced calcification in modern Southern Ocean planktonic foraminifera.* Nature Geoscience 2, 276 – 280 (2009) Published online: 8 March 2009 | doi:10.1038/ngeo460

[53] Turley C, Blackford J, Widdicombe S, Lowe D, Nightingale PD, Rees AP (2006) Reviewing the impact of increased atmospheric CO_2 on oceanic pH and the marine ecosystem. In: Schellnhuber HJ, Cramer W, Nakicenovic N, Wigley T, Yohe G (eds) Avoiding dangerous climate change. Cambridge University Press, Cambridge, pp 65–70

[54] Royal Society (2005) *Ocean acidification due to increasing atmospheric carbon dioxide* Policy document 12/05, Cardiff.

[55] Nature Feature: The New North, 13 October 2011, Vol 478, *Nature* 173

56 http://ozonewatch.gsfc.nasa.gov/Scripts/big_image.
    php?date=2006-09-24&hem=S

57 Zhang, J. Increasing Antarctic Sea Ice under Warming
    Atmospheric and Oceanic Conditions. In *Journal of Climate*, 2007
    DOI: 10.1175/JCLI4136.1

58 From Lionel Carter, I. N. McCave and Michael J. M. Williams,
    Circulation and Water Masses of the Southern Ocean: A Review.
    In: Fabio Florindo and Martin Siegert, editors: Developments
    in Earth and Environmental Sciences, Vol 8, Antarctic Climate
    Evolution, Fabio Florindo and Martin Siegert. The Netherlands:
    Elsevier, 2009, pp. 85–114.

59 Turner et al (2009) Antarctic Climate Change and the Environment
    Scientific Committee on Antarctic Research, Cambridge, UK

60 Turner et al (2009) Antarctic Climate Change and the Environment
    Scientific Committee on Antarctic Research, Cambridge, UK

61 Nicholls, R.J. Sea-Level Rise and Its Impact on Coastal Zones.
    In *Science* Vol 328, 18 July 2010

62 Turner et al (2009) Antarctic Climate Change and the Environment
    Scientific Committee on Antarctic Research, Cambridge, UK

63 Naish, T et al. Examining Antarctica. In *Geotimes*, October 2007,
    pp 30-33

64 IPCC , 2012: Summary for Policymakers. In: *Managing the Risks of
    Extreme Events and Disasters to Advance Climate Change Adaptation*
    [Field, C.B., V. Barros, T.F. Stocker, D. Qin, D.J. Dokken, K.L. Ebi,
    M.D. Mastrandrea, K.J. Mach, G.-K. Plattner, S.K. Allen, M. Tignor,
    and P.M. Midgley (eds.)]. A Special Report of Working Groups
    I and II of the Intergovernmental Panel on Climate Change.
    Cambridge University Press, Cambridge, UK, and New York, NY,
    USA, pp. 1-19.

65 Munich Re Group (2009) *Topics Geo: Natural catastrophes 2008
    Analyses, assessments, positions*

66 King, D.A. Climate Change Science: Adapt, Mitigate, or Ignore?.
    In *Science* Vol 303, 2004

67 Nunn & Mimura (2007) *Promoting Sustainability on Vulnerable
    Island Coasts: A Case Study Smaller Pacific Islands* in McFadden et al
    (eds) *Managing Coastal Vulnerability*, Elsevier, Oxford

68 Strauss, B. et al (2012) *Surging Seas: Sea level rise, storms & global
    warming's threat to the US coast*. Climate Central, USA

[69] IPCC (2007) *Climate Change 2007: Synthesis Report*

[70] Knights, MacGill, Passey *The sustainability of desalination plants in Australia: is renewable energy the answer?*. http://www.ceem.unsw.edu.au/content/userDocs/OzWaterpaperIMRP_000.pdf

[71] Worm, B et al (2006) Impacts of Biodiversity Loss on Ocean Ecosystem. *Science* 314, 787.

[72] Grooten, M et al (2012) *Living Planet Report 2012: Biodiversity, biocapacity and better choices*. World Wildlife Fund, Switzerland.

[73] Lovell, Clare (2009) *Anybody home? Little response in Pacific floor* The America's Intelligence Wire 22 June 2009: n. pag. Web. 17 Oct. 2009

[74] Boyd, P., LaRoche, J., Gall, M., Frew, R.,McKay, R. M. L., 1999. The role of iron, light and silicate in controlling algal biomass in sub-Antarctic water SE of New Zealand. J. Geophys. Res. 104, 13395–13408

[75] Croxall, J. P., Butchart S. H. M., Lascelles, B., Stattersfield A. J., Sullivan B., Symes, A. and Taylor, P. (2012) Seabird conservation status, threats and priority actions: a global assessment. *Bird Conserv. Int.* 22: 1–34.

[76] Peat, N (2003) *Subantarctic New Zealand: A Rare Heritage*. Department of Conservation, Invercargill.

[77] Peat, N (2003) *Subantarctic New Zealand: A Rare Heritage*. Department of Conservation, Invercargill.

[78] SCAR (2005) *Biodiversity in the Antarctic*. Paper prepared for Antarctic Treaty Consultative Meeting 28, IP85.

[79] A.J. & Poloczanska E.S. (2008) Under-Resourced, Under Threat. *Science* 6 June 2008, Vol 320

[80] Bryson, B (2003) *A Short History of Nearly Everything*. Black Swan, London.

[81] Croxall, J. P., Butchart S. H. M., Lascelles, B., Stattersfield A. J., Sullivan B., Symes, A. and Taylor, P. (2012) Seabird conservation status, threats and priority actions: a global assessment. *Bird Conserv. Int.* 22: 1–34.

[82] Taylor, Rowland H. and Taylor, Graeme A.(1989) 'Re-assessment of the status of southern elephant seals (Mirounga leonina) in New Zealand', *New Zealand Journal of Marine and Freshwater Research*, 23: 2, 201 — 213

[83] McMahon, C.R. *et al.* Climate change and seal survival: evidence for environmentally mediated changes in elephant seal, Mirounga leonina , pup survival. Proc. R. Soc. B (2005) 272 , 923–928

[84] Katija, K. & Dabiri, J.O. A viscosity-enhanced mechanism for biogenic ocean mixing. *Nature* 460, 624-626 (30 July 2009) | doi:10.1038/nature08207

[85] Lavery, T.J et al. Iron defecation by sperm whales stimulates carbon export in the Southern Ocean. Proc. R. Soc. B (2010) 277, 3527–3531

[86] Croxall, J. P., Butchart S. H. M., Lascelles, B., Stattersfield A. J., Sullivan B., Symes, A. and Taylor, P. (2012) Seabird conservation status, threats and priority actions: a global assessment. *Bird Conserv. Int.* 22: 1–34.

[87] Yvan Richard, Edward R Abraham & Dominique Filippi (2011). Assessment of the risk to seabird populations from New Zealand commercial fisheries. Final Research Report for Ministry of Fisheries projects IPA2009/19 and IPA2009/20 (Unpublished report held by the Ministry of Fisheries, Wellington). 66 pages.

[88] Robertson, B.C. & Chilvers, B. L. The population decline of the New Zealand sea lion Phocarctos hookeri: a review of possible causes. *Mammal Rev.* 2011

[89] Chilvers, B.L. Population viability analysis of New Zealand sea lions, Auckland Islands, New Zealand's subantarctics: assessing relative impacts and uncertainty. *Polar Biol* (2011) DOI 10.1007/s00300-011-1143-6

[90] Viktoria Kahui. A bioeconomic model for Hooker's sea lion bycatch in New Zealand *Australian Journal of Agricultural and Resource Economics* Volume 56, Issue 1, pages 22–41, January 2012. DOI: 10.1111/j.1467-8489.2011.00566.x

[91] Peat, N (2003) *Subantarctic New Zealand: A Rare Heritage.* Department of Conservation, Invercargill.

[92] SCAR (2005) *Biodiversity in the Antarctic.* Paper prepared for Antarctic Treaty Consultative Meeting 28, IP85.

[93] Searle, J.B. et al. The diverse origins of New Zealand house mice. *Proc. R. Soc. B* (2009) 276, 209–217 doi:10.1098/rspb.2008.0959

[94] Russell, J.C. Spatio-temporal patterns of introduced mice and invertebrates on Antipodes Island. In *Polar Biology* DOI 10.1007/s00300-012-1165-8

95 http://www.acap.aq/latest-news/from-enderby-to-the-antipodes-eradication-of-house-mice-on-southern-ocean-islands

96 BirdLife International 2010. *Procellaria cinerea*. In: IUCN 2011. IUCN Red List of Threatened Species. Version 2011.2. www.iucnredlist.org. Downloaded on 17 April 2012.

97 Thompson, D & Sagar, P. Declining rock hopper penguin populations in New Zealand. *Water & Atmosphere* 10(3) 2002, NIWA

98 McClintock, J., H. Ducklow, and W. Fraser. 2008. Ecological Responses to Climate Change on the Antarctic Peninsula. *American Scientist* 96:302-310.

99 Atkinson, A et al. Long-term decline in krill stock and increase in salps within the Southern Ocean. *Nature* 432, 100-103 (4 November 2004) | doi:10.1038/nature02996;

100 Greg, W. W. et al Ocean primary production and climate: Global decadal changes. In *Geophysical Research Letters*, VOL. 30, NO. 15, 1809, doi:10.1029/2003GL016889, 2003

101 BirdLife International 2010. *Diomedea antipodensis*. In: IUCN 2011. IUCN Red List of Threatened Species. Version 2011.2. www.iucnredlist.org. Downloaded on 17 April 2012.

102 Yvan Richard, Edward R Abraham & Dominique Filippi (2011). Assessment of the risk to seabird populations from New Zealand commercial fisheries. Final Research Report for Ministry of Fisheries projects IPA2009/19 and IPA2009/20 (Unpublished report held by the Ministry of Fisheries, Wellington). 66 pages.

103 Johnson, A. *et al*. Fishing Gear Involved in Entanglements of Right and Humpback Whales. *Marine Mammal Science*, 21(4):635–645 (October 2005)

104 Eriksson, C. & Burton, H. *Collections (1991 to 2001) of marine debris on Maquarie Island show increases in plastic and some fisheries sourced materials* Poster from Australian Antarctic Division, Tasmania.

105 http://www.southernseabirds.org/

106 Benjamin S. Halpern, *et al*. (2008) A Global Map of Human Impact on Marine Ecosystems; DOI: 10.1126/science.1149345 *Science* 319, 948 (2008)

OTHER TITLES BY GARETH MORGAN AND COLLEAGUES

Public Policy Issues
- A Happy Feat (with John McCrystal) (2012) — a voyage to the end of Our Far South
- Hook, Line and Blinkers (with Geoff Simmons) (2011) — everything Kiwis never wanted to know about fishing
- The Big Kahuna (with Susan Guthrie) (2011) — turning tax and welfare in New Zealand on its head
- Poles Apart (with John McCrystal) (2009) — beyond the shouting who's right about climate change?
- Health Cheque (with Geoff Simmons) (2009) — the truth we should all know about New Zealand's public health system
- After the Panic (2009) — surviving bad investments and bad advice
- KiwiSafer (2007) — how to keep your money safe in KiwiSaver
- Pension Panic (2006) — tough talk on sorting your finances.

Motorcycle Travel (all with Jo Morgan)
- Across the Amazon (2011)
- Up the Andes (2010) — across the Andes, north to south, top to bottom
- Under African Skies (2008) — Capetown to London
- Backblocks America (2007) — the US, Canada and Mexico
- Silkriders (2006) — on the trail of Marco Polo.